Communications
in Computer and Information Science **1110**

Commenced Publication in 2007
Founding and Former Series Editors:
Phoebe Chen, Alfredo Cuzzocrea, Xiaoyong Du, Orhun Kara, Ting Liu,
Krishna M. Sivalingam, Dominik Ślęzak, Takashi Washio, Xiaokang Yang,
and Junsong Yuan

Editorial Board Members

Sonja Gievska · Gjorgji Madjarov (Eds.)

ICT Innovations 2019

Big Data Processing and Mining

11th International Conference, ICT Innovations 2019
Ohrid, North Macedonia, October 17–19, 2019
Proceedings

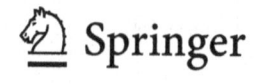 Springer

Editors
Sonja Gievska
Saints Cyril and Methodius
University of Skopje
Skopje, North Macedonia

Gjorgji Madjarov
Saints Cyril and Methodius
University of Skopje
Skopje, North Macedonia

ISSN 1865-0929   ISSN 1865-0937 (electronic)
Communications in Computer and Information Science
ISBN 978-3-030-33109-2   ISBN 978-3-030-33110-8 (eBook)
https://doi.org/10.1007/978-3-030-33110-8

This Springer imprint is published by the registered company Springer Nature Switzerland AG
The registered company address is: Gewerbestrasse 11, 6330 Cham, Switzerland

Preface

The ICT Innovations conference series, organized by the Macedonian Society of Information and Communication Technologies (ICT-ACT) is an international forum for presenting scientific results related to innovative fundamental and applied research in ICT. The 11th ICT Innovations 2019 conference that brought together academics, students, and industrial practitioners, was held in Ohrid, Republic of North Macedonia, during October 17–19, 2019.

The focal point for this year's conference was "Big Data Processing and Mining," with topics extending across several fields including social network analysis, natural language processing, deep learning, sensor network analysis, bioinformatics, FinTech, privacy, and security.

Big data is heralded as one of the most exciting challenges in data science, as well as the next frontier of innovations. The spread of smart, ubiquitous computing and social networking have brought to light more information to consider. Storage, integration, processing, and analysis of massive quantities of data pose significant challenges that have yet to be fully addressed. Extracting patterns from big data provides exciting new fronts for behavioral analytics, predictive and prescriptive modeling, and knowledge discovery. By leveraging the advances in deep learning, stream analytics, large-scale graph analysis and distributed data mining, a number of tasks in fields like, biology, games, robotics, commerce, transportation, and health care have been brought within reach.

Some of these topics were brought to the forefront of the ICT Innovations 2019 conference. This book presents a selection of papers presented at the conference which contributed to the discussions on various aspects of big data mining (including algorithms, models, systems, and applications). The conference gathered 184 authors from 24 countries reporting their scientific work and solutions in ICT. Only 18 papers were selected for this edition by the international Program Committee, consisting of 176 members from 43 countries, chosen for their scientific excellence in their specific fields.

We would like to express our sincere gratitude to the authors for sharing their most recent research, practical solutions, and experiences, allowing us to contribute to the discussion on the trends, opportunities, and challenges in the field of big data. We are grateful to the reviewers for the dedicated support they provided to our thorough reviewing process. Our work was made easier by following the procedures developed and passed along by Prof. Slobodan Kaljadziski, the co-chair of the ICT Innovations 2018 conference. Special thanks to Ilinka Ivanoska, Bojana Koteska, and Monika Simjanoska for their support in organizing the conference and for the technical preparation of the conference proceedings.

October 2019

Sonja Gievska
Gjorgji Madjarov

Organization

Conference and Program Chairs

Sonja Gievska	University Ss.Cyril and Methodius, North Macedonia
Gjorgji Madjarov	University Ss.Cyril and Methodius, North Macedonia

Program Committee

Jugoslav Achkoski	General Mihailo Apostolski Military Academy, North Macedonia
Nevena Ackovska	University Ss.Cyril and Methodius, North Macedonia
Syed Ahsan	Technische Universität Graz, Austria
Marco Aiello	University of Groningen, The Netherlands
Azir Aliu	Southeastern European University of North Macedonia, North Macedonia
Luis Alvarez Sabucedo	Universidade de Vigo, Spain
Ljupcho Antovski	University Ss.Cyril and Methodius, North Macedonia
Stulman Ariel	The Jerusalem College of Technology, Israel
Goce Armenski	University Ss.Cyril and Methodius, North Macedonia
Hrachya Astsatryan	National Academy of Sciences of Armenia, Armenia
Tsonka Baicheva	Bulgarian Academy of Science, Bulgaria
Verica Bakeva	University Ss.Cyril and Methodius, North Macedonia
Antun Balaz	Institute of Physics Belgrade, Serbia
Lasko Basnarkov	University Ss.Cyril and Methodius, North Macedonia
Slobodan Bojanic	Universidad Politécnica de Madrid, Spain
Erik Bongcam-Rudloff	SLU-Global Bioinformatics Centre, Sweden
Singh Brajesh Kumar	RBS College, India
Torsten Braun	University of Berne, Switzerland
Andrej Brodnik	University of Ljubljana, Slovenia
Francesc Burrull	Universidad Politécnica de Cartagena, Spain
Neville Calleja	University of Malta, Malta
Valderrama Carlos	UMons University of Mons, Belgium
Ivan Chorbev	University Ss.Cyril and Methodius, North Macedonia
Ioanna Chouvarda	Aristotle University of Thessaloniki, Greece
Trefois Christophe	University of Luxembourg, Luxembourg
Betim Cico	Epoka University, Albania
Emmanuel Conchon	Institut de Recherche en Informatique de Toulouse, France
Robertas Damasevicius	Kaunas University of Technology, Lithuania
Pasqua D'Ambra	IAC, CNR, Italy
Danco Davcev	University Ss.Cyril and Methodius, North Macedonia
Antonio De Nicola	ENEA, Italy

Slobodan Kalajdziski   University Ss.Cyril and Methodius, North Macedonia
Kalinka Kaloyanova   FMI-University of Sofia, Bulgaria
Aneta Karaivanova   Bulgarian Academy of Sciences, Bulgaria
Mirjana Kljajic Borstnar   University of Maribor, Slovenia
Ljupcho Kocarev   University Ss.Cyril and Methodius, North Macedonia
Dragi Kocev   Jožef Stefan Institute, Slovenia
Margita Kon-Popovska   University Ss.Cyril and Methodius, North Macedonia
Magdalena Kostoska   University Ss.Cyril and Methodius, North Macedonia
Bojana Koteska   University Ss.Cyril and Methodius, North Macedonia
Ivan Kraljevski   VoiceINTERconnect GmbH, Germany
Andrea Kulakov   University Ss.Cyril and Methodius, North Macedonia
Arianit Kurti   Linnaeus University, Sweden
Xu Lai   Bournemouth University, UK
Petre Lameski   University Ss.Cyril and Methodius, North Macedonia
Suzana Loshkovska   University Ss.Cyril and Methodius, North Macedonia
José Machado Da Silva   University of Porto, Portugal
Ana Madevska Bogdanova   University Ss.Cyril and Methodius, North Macedonia
Gjorgji Madjarov   University Ss.Cyril and Methodius, North Macedonia
Tudruj Marek   Polish Academy of Sciences, Poland
Ninoslav Marina   University St. Paul the Apostole, North Macedonia
Smile Markovski   University Ss.Cyril and Methodius, North Macedonia
Marcin Michalak   Silesian University of Technology, Poland
Hristina Mihajlovska   University Ss.Cyril and Methodius, North Macedonia
Marija Mihova   University Ss.Cyril and Methodius, North Macedonia
Aleksandra Mileva   University Goce Delcev, North Macedonia
Biljana Mileva Boshkoska   Faculty of Information Studies in Novo Mesto,
                                Slovenia
Georgina Mirceva   University Ss.Cyril and Methodius, North Macedonia
Miroslav Mirchev   University Ss.Cyril and Methodius, North Macedonia
Igor Mishkovski   University Ss.Cyril and Methodius, North Macedonia
Kosta Mitreski   University Ss.Cyril and Methodius, North Macedonia
Pece Mitrevski   University St. Kliment Ohridski, North Macedonia
Irina Mocanu   University Politehnica of Bucharest, Romania
Ammar Mohammed   Cairo University, Egypt
Andreja Naumoski   University Ss.Cyril and Methodius, North Macedonia
Manuel Noguera   University of Granada, Spain
Thiare Ousmane   Gaston Berger University, Senegal
Eleni Papakonstantinou   Genetics Lab, Greece
Marcin Paprzycki   Polish Academy of Sciences, Poland
Dana Petcu   West University of Timisoara, Romania
Antonio Pinheiro   Universidade da Beira Interior, Portugal
Matus Pleva   Technical University of Košice, Slovakia
Florin Pop   University Politehnica of Bucharest, Romania
Zaneta Popeska   University Ss.Cyril and Methodius, North Macedonia
Aleksandra   University Ss.Cyril and Methodius, North Macedonia
  Popovska-Mitrovikj

Marco Porta	University of Pavia, Italy
Ustijana Rechkoska Shikoska	University of Information Science and Technology St. Paul The Apostle, North Macedonia
Manjeet Rege	University of St. Thomas, USA
Bernd Rinn	ETH Zurich, Switzerland
Blagoj Ristevski	University St. Kliment Ohridski, North Macedonia
Sasko Ristov	University Ss.Cyril and Methodius, North Macedonia
Witold Rudnicki	University of Białystok, Poland
Jelena Ruzic	Mediteranean Institute for Life Sciences, Croatia
David Šafránek	Masaryk University, Czech Republic
Simona Samardjiska	University Ss.Cyril and Methodius, North Macedonia
Wibowo Santoso	Central Queensland University, Australia
Snezana Savovska	University St. Kliment Ohridski, North Macedonia
Loren Schwiebert	Wayne State University, USA
Vladimir Siládi	Matej Bel University, Slovakia
Josep Silva	Universitat Politècnica de València, Spain
Ana Sokolova	University of Salzburg, Austria
Michael Sonntag	Johannes Kepler University Linz, Austria
Dejan Spasov	University Ss.Cyril and Methodius, North Macedonia
Todorova Stela	University of Agriculture, Bulgaria
Goran Stojanovski	Elevate Global, North Macedonia
Biljana Stojkoska	University Ss.Cyril and Methodius, North Macedonia
Ariel Stulman	The Jerusalem College of Technology, Israel
Spinsante Susanna	Università Politecnica delle Marche, Italy
Ousmane Thiare	Gaston Berger University, Senegal
Biljana Tojtovska	University Ss.Cyril and Methodius, North Macedonia
Yalcin Tolga	NXP Labs, UK
Dimitar Trajanov	University Ss.Cyril and Methodius, North Macedonia
Ljiljana Trajkovic	Simon Fraser University, Canada
Vladimir Trajkovik	University Ss.Cyril and Methodius, North Macedonia
Denis Trcek	University of Ljubljana, Slovenia
Christophe Trefois	University of Luxembourg, Luxembourg
Kire Trivodaliev	University Ss.Cyril and Methodius, North Macedonia
Katarina Trojacanec	University Ss.Cyril and Methodius, North Macedonia
Hieu Trung Huynh	Industrial University of Ho Chi Minh City, Vietnam
Zlatko Varbanov	Veliko Tarnovo University, Bulgaria
Goran Velinov	University Ss.Cyril and Methodius, North Macedonia
Elena Vlahu-Gjorgievska	University of Wollongong, Australia
Irena Vodenska	Boston University, USA
Katarzyna Wac	University of Geneva, Switzerland
Yue Wuyi	Konan University, Japan
Zeng Xiangyan	Fort Valley State University, USA
Shuxiang Xu	University of Tasmania, Australia
Rita Yi Man Li	Hong Kong Shue Yan University, Hong Kong, China
Malik Yousef	Zefat Academic College, Israel

Massimiliano Zannin	INNAXIS Foundation Research Institute, Spain
Zoran Zdravev	University Goce Delcev, North Macedonia
Eftim Zdravevski	University Ss.Cyril and Methodius, North Macedonia
Vladimir Zdravevski	University Ss.Cyril and Methodius, North Macedonia
Katerina Zdravkova	University Ss.Cyril and Methodius, North Macedonia
Jurica Zucko	Faculty of Food Technology and Biotechnology, Croatia
Chang Ai Sun	University of Science and Technology Beijing, China
Yin Fu Huang	University of Science and Technology, Taiwan
Suliman Mohamed Fati	INTI International University, Malaysia
Hwee San Lim	University Sains Malaysia, Malaysia
Fu Shiung Hsieh	University of Technology, Taiwan
Zlatko Varbanov	Veliko Tarnovo University, Bulgaria
Dimitrios Vlachakis	Genetics Lab, Greece
Boris Vrdoljak	University of Zagreb, Croatia
Maja Zagorščak	National Institute of Biology, Slovenia

Scientific Committee

Danco Davcev	University Ss.Cyril and Methodius, North Macedonia
Dejan Gjorgjevikj	University Ss.Cyril and Methodius, North Macedonia
Boro Jakimovski	University Ss.Cyril and Methodius, North Macedonia
Aleksandra Popovska-Mitrovikj	University Ss.Cyril and Methodius, North Macedonia
Sonja Gievska	University Ss.Cyril and Methodius, North Macedonia
Gjorgji Madjarov	University Ss.Cyril and Methodius, North Macedonia

Technical Committee

Ilinka Ivanoska	University Ss.Cyril and Methodius, North Macedonia
Monika Simjanoska	University Ss.Cyril and Methodius, North Macedonia
Bojana Koteska	University Ss.Cyril and Methodius, North Macedonia
Martina Toshevska	University Ss.Cyril and Methodius, North Macedonia
Frosina Stojanovska	University Ss.Cyril and Methodius, North Macedonia

Additional Reviewers

Emanouil Atanassov
Emanuele Pio Barracchia

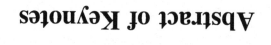

Abstract of Keynotes

Machine Learning Optimization and Modeling: Challenges and Solutions to Data Deluge

Diego Klabjan[1,2,3]

[1] Department of Industrial Engineering and Management Sciences,
Northwestern University
[2] Master of Science in Analytics, Northwestern University
[3] Center for Deep Learning, Northwestern University
d-klabjan@northwestern.edu

Abstract. A single server can no longer handle all of the data of a machine learning problem. Today's data is fine granular, usually has the temporal dimension, is often streamed, and thus distributed among several compute nodes on premise or in the cloud. More hardware buys you only so much; in particular, the underlying models and algorithms must be capable of exploiting it. We focus on distributed optimization algorithms where samples and features are distributed, and in a different setting where data is streamed by an infinite pipeline. Algorithms and convergence analyses will be presented. Fine granular data with a time dimension also offers opportunities to deep learning models that outperform traditional machine learning models. To this end, we use churn predictions to showcase how recurrent neural networks with several important enhancements squeeze additional business value.

Keywords: Distributed optimization • Deep learning • Recurrent neural networks

Computing and Probing Cancer Immunity

Zlatko Trajanoski

Division of Bioinformatics, Medical University of Innsbruck
zlatko.trajanoski@i-med.ac.at

Abstract. Recent breakthroughs in cancer immunotherapy and decreasing costs of high-throughput technologies sparked intensive research into tumour-immune cell interactions using genomic tools. However, the wealth of the generated data and the added complexity pose considerable challenges and require computational tools to process, analyse and visualise the data. Recently, a number of tools have been developed and used to effectively mine tumour immunologic and genomic data and provide novel mechanistic insights. In this talk I will first review and discuss computational genomics tools for mining cancer genomic data and extracting immunological parameters. I will focus on higher-level analyses of NGS data including quantification of tumour-infiltrating lymphocytes (TILs), identification of tumour antigens and T cell receptor (TCR) profiling. Additionally, I will address the major challenges in the field and ongoing efforts to tackle them.

In the second part I will show results generated using state-of-the-art computational tools addressing several prevailing questions in cancer immunology including: estimation of the TIL landscape, identification of determinants of tumour immunogenicity, and immuno editing that tumors undergo during progression or as a consequence of targeting the PD-1/PD-L1 axis. Finally, I will propose a novel approach based on perturbation biology of patient-derived organoids and mathematical modeling for the identification of a mechanistic rationale for combination immunotherapies in colorectal cancer.

Keywords: Cancer immunotherapy • Tumour-infiltrating lymphocytes • Perturbation biology

Bioinformatics Approaches for Sing Cell Transcriptomics and Big Omics Data Analysis

Ming Chen

Department of Bioinformatics, College of Life Sciences, Zhejiang University
mchen@zju.edu.cn

Abstract. We are in the big data era. Multi-omics data brings us a challenge to develop appropriate bioinformatics approaches to model complex biological systems at spatial and temporal scales. In this talk, we will describe multi-omics data available for biological interactome modeling. Single cell transcriptomics data is exploited and analyzed. An integrative interactome model of non-coding RNAs is built. We investigated to characterize coding and non-coding RNAs including microRNAs, siRNAs, lncRNAs, ceRNAs and cirRNAs.

Keywords: Big data • Multi-omics data • RNA

Crosslingual Document Embedding
as Reduced-Rank Ridge Regression

Robert West

Tenure Track Assistant Professor, Data Science Laboratory, EPFL
robert.west@epfl.chs

Abstract. There has recently been much interest in extending vector-based word representations to multiple languages, such that words can be compared across languages. In this paper, we shift the focus from words to documents and introduce a method for embedding documents written in any language into a single, language-independent vector space. For training, our approach leverages a multilingual corpus where the same concept is covered in multiple languages (but not necessarily via exact translations), such as Wikipedia. Our method, Cr5 (Crosslingual reduced-rank ridge regression), starts by training a ridge-regression-based classifier that uses language-specific bag-of-word features in order to predict the concept that a given document is about. We show that, when constraining the learned weight matrix to be of low rank, it can be factored to obtain the desired mappings from language-specific bags-of-words to language-independent embeddings. As opposed to most prior methods, which use pretrained monolingual word vectors, postprocess them to make them crosslingual, and finally average word vectors to obtain document vectors, Cr5 is trained end-to-end and is thus natively crosslingual as well as document-level. Moreover, since our algorithm uses the singular value decomposition as its core operation, it is highly scalable. Experiments show that our method achieves state-of-the-art performance on a crosslingual document retrieval task. Finally, although not trained for embedding sentences and words, it also achieves competitive performance on crosslingual sentence and word retrieval tasks.

Keywords: Crosslingual · Reduced-rank · Ridge regression ·
Retrieval · Embeddings

Contents

Automatic Text Generation in Macedonian Using Recurrent Neural Networks

Ivona Milanova, Ksenija Sarvanoska, Viktor Srbinoski,
and Hristijan Gjoreski[✉]

Faculty of Electrical Engineering and Information Technologies,
University of Ss. Cyril and Methodius in Skopje, Skopje, North Macedonia
ivonamilanova221@gmail.com,
ksenija.sarvanoska@gmail.com,
viktor_srbinoski@hotmail.com,
hristijang@feit.ukim.edu.mk

Abstract. Neural text generation is the process of a training neural network to generate a human understandable text (poem, story, article). Recurrent Neural Networks and Long-Short Term Memory are powerful sequence models that are suitable for this kind of task. In this paper, we have developed two types of language models, one generating news articles and the other generating poems in Macedonian language. We developed and tested several different model architectures, among which we also tried transfer-learning model, since text generation requires a lot of processing time. As evaluation metric we used ROUGE-N metric (Recall-Oriented Understudy for Gisting Evaluation), where the generated text was tested against a reference text written by an expert. The results showed that even though the generate text had flaws, it was human understandable, and it was consistent throughout the sentences. To the best of our knowledge this is a first attempt in automatic text generation (poems and articles) in Macedonian language using Deep Learning.

Keywords: Text generation · Storytelling · Poems · RNN · Macedonian language · NLP · Transfer learning · ROUGE-N

1 Introduction

As the presence of Artificial Intelligence (AI) and Deep Learning has become more prominent in the past couple of years and the fields have acquired significant popularity, more and more tasks from the domain of Natural Language Processing are being implemented. One such task is automatic text generation, which can be designed with the help of deep neural networks, especially Recurrent Neural Networks [16]. Text generation is the process of preparing text for developing a word-based language model and designing and fitting a neural language model in such a way that it can predict the likelihood of occurrence of a word based on the previous sequence of words used in the source text. After that the learned language model is used to generate new text with similar statistical properties as the source text.

S. Gievska and G. Madjarov (Eds.): ICT Innovations 2019, CCIS 1110, pp. 1–12, 2019.
https://doi.org/10.1007/978-3-030-33110-8_1

In our paper, we do an experimental evaluation of two types of word-based language models in order to create a text generation system that will generate text in Macedonian. The first model is generating paragraph of a news article and the other is generating poems. For generating news articles, we also tried implementing transfer learning, but as there are no pre-trained models on a dataset in Macedonian, we used a model that was trained on a dataset in English, so the results were not satisfying. The second model we created was used to generate poetry and was trained on a dataset consisting of Macedonian folk poems. In order to measure how closely the generated text resembles a human written text, we used a metric called ROUGE-N (Recall-Oriented Understudy for Gisting Evaluation), which is a set of metrics for evaluating automatic generation of texts as well as machine translation. With this metric, we got an F1 score of 65%.

2 Related Work

Recent years have brought a huge interest in language modeling tasks, a lot of them being automatic text generation from a corpus of text, as well as visual captioning and video summarization. The burst of deep learning and the massive development of hardware infrastructure has made this task much more possible.

Some of the tasks in this field include automatic text generation based on intuitive model and using heuristics to look for the elements of the text that were proposed by human feedback [1, 2]. Another approach has leaned towards character-based text generation using a Hessian-free optimization in order to overcome the difficulties associated with training RNNs [3]. Text generation using independent short description has also been one topic of research. The purpose of this paper has been to describe a scene or event using independent descriptions. They have used both Statistical Machine Translation and Deep Learning to present text generation in two different manners [4]. The other kind of text generation application has been designing text based interactive narratives. Some of them have been using an evolutionary algorithm with an end-to-end system that understands the components of the text generation pipeline stochastically [5] and others have been using mining of crowd sourced information from the web [6, 7].

Many papers have also focused on visual text generation, image captioning and video description. One recent approach to image captioning used CNN-LSTM structures [8, 9]. Sequence-to-sequence models have been used to caption video or movie contents. They are using an approach where the first sequence encodes the video and the second decodes the description [10, 11].

The idea behind document summarization has been used on video summarization where instead of extracting key sentences, key frames or shots are selected [12].

Visual storytelling is the process of telling a coherent story about an image set. Some of the works covering this include storyline graph modeling [13] and unsupervised mining [14].

Another state-of-the-art approach includes using hierarchically structured reinforcement learning for generating coherent multi-sentence stories for the visual storytelling task [25].

However, all of these approaches include text generation in English where the amount of available data is enormous. Our paper focuses on creating stories in Macedonian using data from Macedonian news portals and folk poetry as well as exploration of different model architectures in order to get the best result, regarding comprehensibility and execution time.

3 Dataset and Preprocessing

The first dataset that we used was gathered from online news portals and consisted of around 2.5 million words. The data was collected with the help of a scraper program we wrote in .NET Core using the C# programming language. The program loads the web page from a given Uniform Resource Locator (URL) and then looks at the html tags. When it finds a match for the html tags that we have given the program, it takes their content and writes it to a file.

The second dataset consisted of a collection of Macedonian poetry written by various Macedonian writers [17] and was made up of roughly 7 thousand words. The data was not clean so we had to do a fair amount of data preprocessing.

The collected datasets are publically available at https://github.com/Ivona221/ MacedonianStoryTelling.

In order to prepare the data that we collected to be suitable to enter the algorithm and to simplify the task of the algorithm when it starts learning, we had to do a considerable amount of data cleaning. For this purpose, we created a pipeline in C#, which in the end gave us a clean dataset to work on. The first step in the algorithm was to remove any special characters from the text corpus including html tags that were extracted from the websites along with the text, JavaScript functions and so on. We also had to translate some of the symbols into text like dollar signs, degree signs and mathematical operators in order to have the least amount of unique characters for the algorithm to work with. The next step was to translate all the English words if there existed a translation or remove the sentences where that was not the case. Next, we had to separate all the punctuation signs from the words with an empty space in order for the algorithm to consider them as independent words. The last and one of the most important steps in this pipeline was creating a custom word tokenizer. All the existing word tokenizers were making a split on an empty space. However, they do not take into consideration the most common word collocations as well as words containing dash, name initials and abbreviations. Our algorithm was taking these words as one. The abbreviations were handled by using a look-up table of all the Macedonian abbreviations, as well as the most common collocations and the initials were handled by searching for specific patterns in text like capital letter-point-capital letter (Fig. 1).

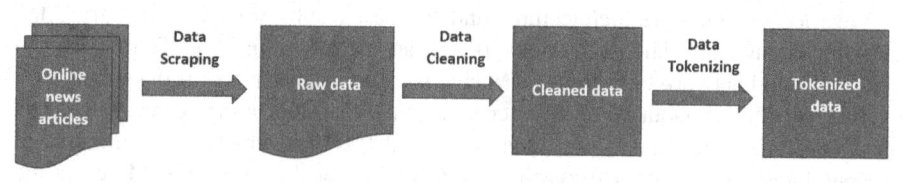

Fig. 1. Data preprocessing flow

4   Language Model Architecture

We have trained two types of models that work on the principle of predicting the next word in a sequence, one for news generation and one for poem generation. The language models used were statistical and predicted the probability of each word, given an input sequence of text. For the news generation model, we created several different variations, including a transfer learning approach (Fig. 2).

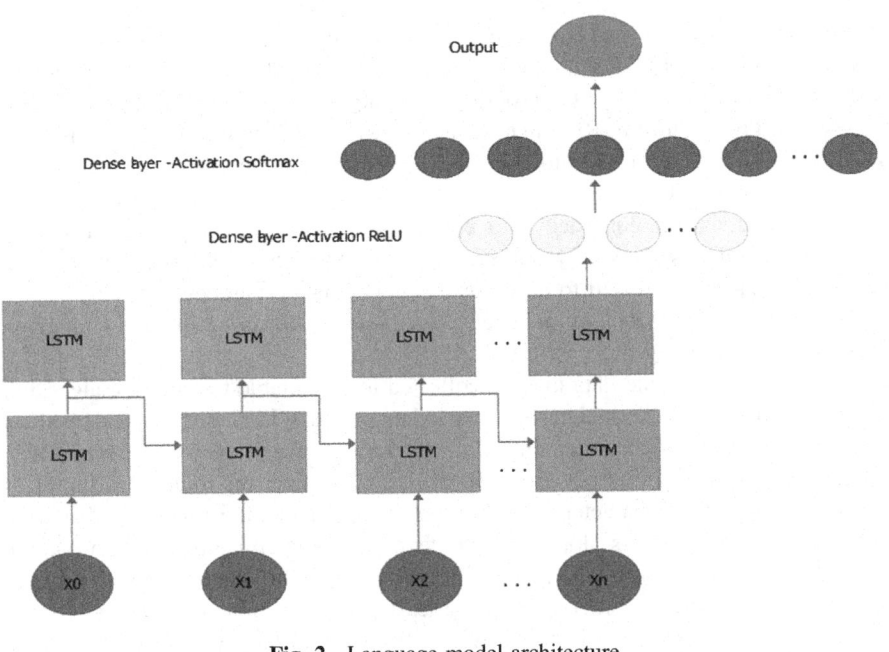

Fig. 2. Language model architecture

4.1   News Article Generation

The first approach for news generation was trained on news articles and used a sequence of hundred words as input. It then generates one word based on that sequence, meaning the word with the biggest probability of appearing next. In the next time step it adds the generated word to the sequence of a hundred words and cuts out the very first word, meaning that it once again makes a sequence with a length of one hundred. It then feeds this new sequence to itself as input for the next time step and continues doing so until it generates the preset amount of words.

We tried out multiple architectures and the best results were acquired from the following architecture. The neural network was an LSTM (Long-Short Term Memory) recurrent neural network with two LSTM layers and two dense layers and also, we tried a variation with a dropout layer in order to see how that affects the performance. The first LSTM layer consists of 100 hidden units and 100 timesteps and is configured to give one hidden state output for each input time step for the single LSTM cell in the

layer. The second layer we added was a Dropout layer with a dropout rate of 0.2. They key idea behind adding a dropout layer is to prevent overfitting. This technique works by randomly dropping units (along with their connections) from the neural network during training. This prevents units from co-adapting too much [18]. The next layer is also an LSTM layer with 100 hidden units. Then we added a Dense layer which is a fully connected layer. A dense layer represents a matrix vector multiplication. The values in the matrix are the trainable parameters, which get updated during back-propagation. Assuming we have an n-dimensional input vector u (in our case 100-dimensional input vector) presented with the formula:

$$u \in R^{n*1} \tag{1}$$

We get an m-dimensional vector as output.

$$u^{T.W} = W \in R^{n*m} \tag{2}$$

A dense layer thus is used to change the dimensions of the vector. Mathematically speaking, it applies a rotation, scaling, translation transform to your vector. The activation function meaning the element-wise function we applied on this layer was ReLU (Rectified Linear Units). ReLU is an activation function introduced by [19]. In 2011, it was demonstrated to improve training of deep neural networks. It works by thresholding values at zero, i.e. $f(x) = max\ (0, x)$. Simply put, it outputs zero when $x < 0$, and conversely, it outputs a linear function when $x \geq 0$. The last layer was also a Dense layer, however with a different activation function, softmax. Softmax function calculates the probability distribution of the event over 'n' different events. In a general way of saying, this function will calculate the probabilities of each target class over all possible target classes. Later the calculated probabilities will be helpful for determining the target class for the given inputs. As the loss function or the error function we decided to use sparse categorical cross entropy. A loss function compares the predicted label and true label and calculates the loss. With categorical cross entropy, the formula to compute the loss is as follows:

$$-\sum_{c=1}^{M} y_{o,c} \log(p_{o.c}) \tag{3}$$

where,

- M – number of classes
- log – the natural log
- y – binary indicator (0 or 1) if class label c is the correct classification for observation o
- p – predicted probability observation o is of class c

The only difference between sparse categorical cross entropy and categorical cross entropy is the format of true labels. When we have a single-label, multi-class classification problem, the labels are mutually exclusive for each data, meaning each data entry can only belong to one class. Then we can represent y_true using one-hot embeddings. This saves memory when the label is sparse (the number of classes is very large).

As an optimizer we decided to use Adam optimizer (Adaptive Moment Estimation) which is method that computes adaptive learning rates for each parameter. It is an algorithm for first-order gradient-based optimization of stochastic objective functions. In addition to storing an exponentially decaying average of past squared gradients vt like Adadelta and RMSprop, Adam also keeps an exponentially decaying average of past gradients m_t, similar to momentum [20]. As an evaluation metric we used accuracy.

For the second approach, we tried using a transfer learning method using the pre trained model word2vec, which after each epoch generated as many sentences as we gave it starting words on the beginning. Word2vec is a two-layer neural net that processes text. On the pre-trained model, we added one LSTM layer and one Dense layer. When using a pretrained model the embedding or the first layer in our model is seeded with the word2vec word embedding weights. We trained this model on 100 epochs using a batch size of 128. However, as the pre-trained models are trained using English text, this model did not give us satisfying results and that is the reason why this models results will not be observed in this paper [22].

4.2 Poem Generation

This model has only slight differences from the first LSTM network we described. It is an LSTM neural network as well, with two LSTM layers and one dense layer. Having experimented with several different combinations of parameters we decided on the following architecture. The first LSTM layer is made up of 150 units, then we have a Dropout layer with a dropout rate of 0.2 so that we can reduce overfitting. The second LSTM layer is made up of 100 units and it has another Dropout layer after it with a dropout rate of 0.2. At last, we have a dense layer with a softmax activation function, which picks out the most fitting class or rather word for the given input.

In this model we have decided on the loss function to be categorical cross entropy and as an optimizer once again we use the Adam optimizer.

Another thing that we do differently with the second model is the usage of a callback function EarlyStop [21]. A callback is a set of functions that are applied at certain stages of the training procedure with the purpose of viewing the internal states and statistics of the model during training. EarlyStop helps lessen the problem of how long to train a network, since too little training could lead to under fitting to the train and test sets while too much training leads to overfitting. We train the network on a training set until the performance on a validation set starts to degrade. When the model starts learning the statistical noise in the training set and stops generalizing, the generalizing error will increase and signal overfitting. With this approach during the training after every epoch, we evaluate the model on a holdout validation dataset and if this performance starts decaying then the training process is stopped. Because we are certain that the network will stop at an appropriate point in time, we use a large number of training epochs, more than normally required, so that the network is given an opportunity to fit and then begin to over fit to the training set. In our case, we use 100 epochs.

Early stopping is probably the oldest and most widely used form of neural network regularization.

5 Text Generation

As mentioned before the first language model is fed a sequence of hundred words and to make this possible, a few steps in the text preprocessing need to be taken. With the tokenizer we first vectorize the data by turning the text into a sequence of integers, each integer being the index of a token in a dictionary. We then construct a new file which contains our input text but makes sure to have one hundred words per each line. In the text generation process we randomly select one line from the previously created file for the purpose of generating a new word. We then encode this line of text to integers using the same tokenizer that used when training the model. The model then makes a prediction of the next word and gives an index of the word with the highest probability which we must look up in the Tokenizers mapping to retrieve the associated word. We then append this new word to the seed text and repeat the process.

Considering that the sequence will eventually become too long we can truncate it to the appropriate length after the input sequence has been encoded to integers.

6 Results

As we mentioned before we trained two different main models one for news article generation and another one for poems.

The accuracy and loss for this kind of task are calculated on the train set since one cannot measure the correctness of a story, therefore test set cannot be constructed. In order to evaluate the result we compared it against a human produced equivalent of the generated story.

Regarding the news generation model, we tried two variations, one with dropout and one without dropout layers and tested how that affected the training accuracy and loss. Both variations were trained on 50 epochs, using a batch size of 64. Comparing the results, adding dropout layers improved accuracy and required shorter training time (Figs. 3 and 4).

The poem generation model was trained on 100 epochs using a batch size of 64. This model also included dropout layers (Figs. 5 and 6).

In order to evaluate how closely the generated text resembles a human written text, we used ROUGE-N metric. It works by comparing an automatically produced text or translation against a set of reference texts, which are human-produced. The recall (in the context of ROUGE) refers to how much of the reference summary the system summary is recovering or capturing. It can be computed as:

$$\frac{number\_of\_overlapping\_words}{total\_words\_in\_reference\_text} \tag{4}$$

The precision measures how much of the system summary was in fact relevant or needed. It is calculated as:

$$\frac{number\_of\_overlapping\_words}{total\_words\_in\_system\_generated\_text} \tag{5}$$

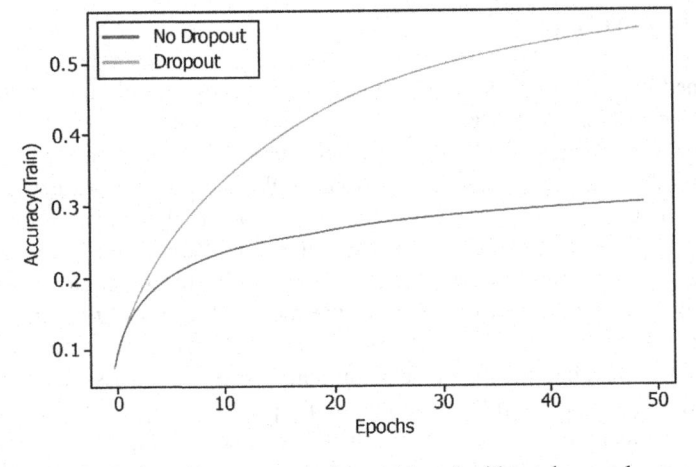

Fig. 3. Train accuracy, comparison with and without dropout layer

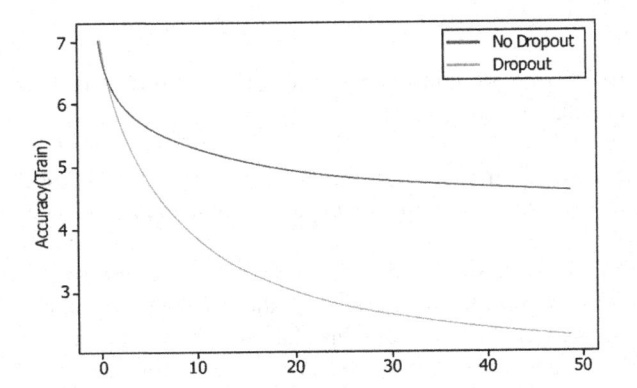

Fig. 4. Train loss, comparison with and without dropout layer

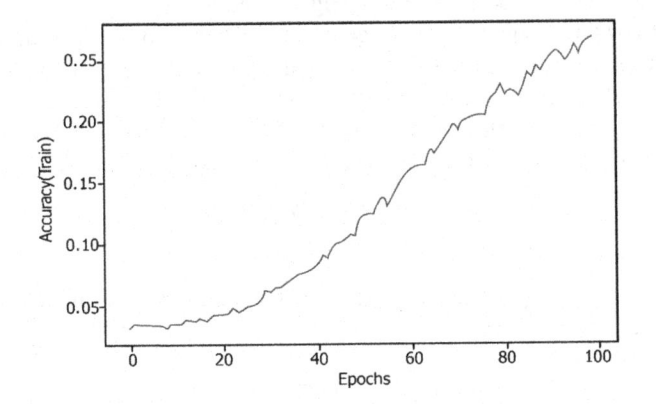

Fig. 5. Train accuracy

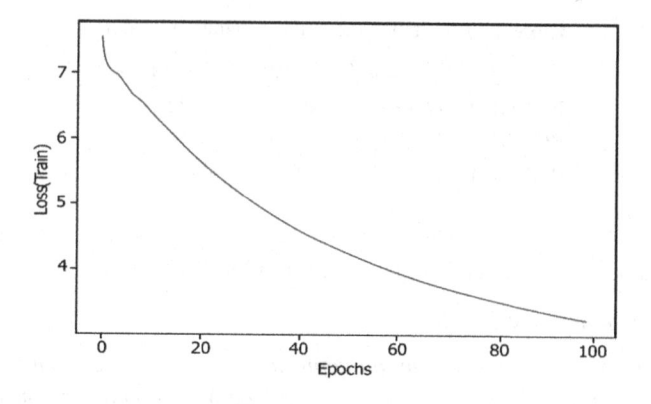

Fig. 6. Train loss.

Using the precision and recall, we can compute an F1 score with the following formula:

$$F1 = \frac{precision * recall}{precision + recall} * 2 \tag{6}$$

In our case we used ROUGE-1, ROUGE-2 and ROUGE-L.

- ROUGE-1 refers to overlap of unigrams between the system summary and reference summary
- ROUGE-2 refers to the overlap of bigrams between the system and reference summaries
- ROUGE-L measures longest matching sequence of words using LCS (Longest Common Subsequence). An advantage of using LCS is that it does not require consecutive matches, but in-sequence matches that reflect sentence level word order. Since it automatically includes longest in-sequence common n-grams, you do not need a predefined n-gram length.

The reason one would use ROUGE-2 over or in conjunction with ROUGE-1, is to also show the fluency of the texts or translations. The intuition is that if you follow the word orderings of the reference summary more closely, then your summary is actually more fluent [24].

The results from the ROUGE-N metric are shown in Tables 1 and 2:

Table 1. Results for the news generation model

	Precision %	Recall %	F1-score %
ROUGE-1	66.67	76.92	71.43
ROUGE-2	35.85	57.58	44.19
ROUGE-L	66.67	76.92	70.71

Table 2. Results for the poem generation model

	Precision %	Recall %	F1-score %
ROUGE-1	47.89	46.58	47.22
ROUGE-2	21.05	21.28	21.16
ROUGE-L	40.85	39.73	40.26

In the end we present a sample generated from each of the models:

- News generation model:

...ќе се случи и да се користат на корисниците, а и да се случи, се наоѓа во играта, а и да се случи, се уште не се случува во Македонија. Во моментов, во текот на државата се во прашање, се во прашање, но не и да се случи, но и за да се случи, се наведува во соопштението на САД. Во текот на државата се во прашање, се во прашање, и да се случи, се уште не и корупциски скандали, се наоѓа во водата и...

- Poem generation model:

Во залула време тамо - од прангите свекот, оро на земниот
земниот роден кат, - трендафил во одаја, плиснала врева. магла
на очиве моја! жетва најбогата цвета, оро на кај, шар.
јазик љубовта наша делата в тих. рака јазик падна, -
стиснат сите луѓе - нок, убава, красна, дише рој, писнат,
најде на туѓина - чемер во одаја, шар. јазик идеш
праг, роден кат, и скитник со здржана туѓи оро син
стиснала - ПОЕМА - нок, убава, красна, дише на робја,
шар. издржа сите луѓе - нок, убава, красна, дише рој,
писнат, мајка, кат, - развива ора, во онаа вечер, в
очи очиве носам, у секоа доба. оро ќе сретнам, најде.

7   Conclusion

In this paper we present a solution to the problem of automatic text generation in Macedonian language. To the best of our knowledge this is a first attempt in automatic text generation (poems and articles) in Macedonian language using Deep Learning.

We made attempts with two types of models, the first for news generation and the second for poem generation. We also tried a transfer learning model using word-2-vec, however the results were not satisfying. Excluding the network where we used transfer learning, we got promising results. The first model was able to generate a text of a hundred words, including punctuation marks. It used syntax rules correctly and put punctuation marks where needed. Even though the generated output of the poem model was nonsensical, it was still a clear indication that the network was able to learn the style of the writers and compose a similar looking piece of work.

There is, of course, a lot left to be desired and a lot more can be done to improve upon our work, such as using a more cohesive dataset. Because our dataset was created by putting different news articles together, there is no logical connection from one news article to the other, which resulted in our model not having an output that made much sense. The same point can apply to our dataset of poems where each poem was standalone and there was no connection from one to the other. Another suggestion for achieving better results is adding more layers and units to the networks, keeping in mind that as the size of the neural network gets bigger, so do the hardware requirements and training time needed [23].

Acknowledgment. We gratefully acknowledge the support of NVIDIA Corporation with the donation of the Titan Xp GPU used for this research.

References

1. Bailey, P.: Searching for storiness: story-generation from a reader's perspective. In: Working Notes of the Narrative Intelligence Symposium (1999)
2. PÉrez, R.P.Ý., Sharples, M.: MEXICA: a computer model of a cognitive account of creative writing. J. Exp. Teor. Artif. Intell. **13**, 119–139 (2001)
3. Sutskever, I., Martens, J., Hinton, G.E.: Generating text with recurrent neural networks. In: Proceedings of the 28th International Conference on Machine Learning (ICML-2011), pp. 1017–1024 (2011)
4. Jain, P., Agrawal, P., Mishra, A., Sukhwani, M., Laha, A., Sankaranarayanan, K.: Story generation from sequence of independent short descriptions. In: Proceedings of Workshop on Machine Learning for Creativity, Halifax, Canada, August 2017 (SIGKDD 2017) (2017)
5. McIntyre, N., Lapata, M.: Learning to tell tales: a data-driven approach to story generation. In: Proceedings of the Joint Conference of the 47th Annual Meeting of the ACL and the 4th International Joint Conference on Natural Language Processing of the AFNLP, Association for Computational Linguistics, vol. 1, pp. 217–225 (2009)
6. Li, B., Lee-Urban, S., Johnston, G., Riedl, M.: Story generation with crowdsourced plot graphs. In: AAAI (2013)
7. Swanson, R., Gordon, A.: Say anything: a massively collaborative open domain story writing companion. Interact. Storytelling **2008**, 32–40 (2008)
8. Vinyals, O., Toshev, A., Bengio, S., Erhan, D.: Show and tell: a neural image caption generator. In: CVPR (2015)
9. Xu, K., et al.: Show, attend and tell: neural image caption generation with visual attention. In: ICML (2015)
10. Venugopalan, S., Rohrbach, M., Donahue, J., Mooney, R., Darrell, T., Saenko, K.: Sequence to sequence-video to text. In: ICCV (2015)
11. Pan, Y., Mei, T., Yao, T., Li, H., Rui, Y.: Jointly modeling embedding and translation to bridge video and language. In: CVPR (2016)
12. Rush, A.M., Chopra, S., Weston, J.: A neural attention model for abstractive sentence summarization. In: EMNLP (2015)
13. Kim, G., Xing, E.P.: Reconstructing storyline graphs for image recommendation from web community photos. In: CVPR (2014)
14. Sigurdsson, G.A., Chen, X., Gupta, A.: Learning visual storylines with skipping recurrent neural networks. In: ECCV (2016)

15. Glorianna Jagfeld, S.J.: Sequence-to-sequence models for data-to-text natural language (2018)
16. Sutskever, I.: Generating text with recurrent neural networks. In: 28th International Conference on Machine Learning (ICML-2011), pp. 1017–1024 (2011)
17. Racin, K.: Beli mugri, pp. 3–33. Makedonska kniga, Skopje (1989)
18. Srivastava, N., Hinton, G., Krizhevsky, A., Sutskever, I., Salakhutdinov, R.: Dropout: a simple way to prevent neural networks from overfitting. J. Mach. Learn. Res. **15**, 1929–1958 (2014)
19. Hahnloser, R.H., Sarpeshkar, R., Mahowald, M.A., Douglas, R.J., Seung, H.S.: Digital selection and analogue amplification coexist in a cortex-inspired silicon circuit. Nature **405** (6789), 947 (2000)
20. Kingma, D.P., Ba, J.: Adam: A Method for Stochastic Optimization. arXiv:1412.6980 [cs. LG], December 2014
21. Montavon, G., Orr, Geneviève B., Müller, K.-R. (eds.): Neural Networks: Tricks of the Trade. LNCS, vol. 7700. Springer, Heidelberg (2012). https://doi.org/10.1007/978-3-642-35289-8
22. Mikolov, T., et al.: Distributed representations of words and phrases and their compositionality. In: NIPS (2013)
23. Pradhan, S.: Exploring the depths of recurrent neural networks with stochastic residual learning (2016)
24. Lin, C.-Y.: ROUGE: a package for automatic evaluation of summaries. In: ACL Workshop: Text Summarization Braches Out 2004, p. 10 (2004)
25. Huang, Q., Gan, Z., Celikyilmaz, A., Wu, D., Wang, J., He, X.: Hierarchically structured reinforcement learning for topically coherent visual story generation. In: Proceedings of the AAAI Conference on Artificial Intelligence, vol. 33, pp. 8465–8472, July 2019

Detection of Toy Soldiers Taken from a Bird's Perspective Using Convolutional Neural Networks

Saša Sambolek and Marina Ivašić-Kos[(✉)]

Department of Informatics, University of Rijeka, Rijeka, Croatia
sasa.sambolek@gmail.com, marinai@inf.uniri.hr

Abstract. This paper describes the use of two different deep-learning approaches for object detection to recognize a toy soldier. We use recordings of toy soldiers in different poses under different scenarios to simulate appearance of persons on footage taken by drones. Recordings from a bird's eye view are today widely used in the search for missing persons in non-urban areas, border control, animal movement control, and the like. We have compared the single-shot multi-box detector (SSD) with the MobileNet or Inception V2 as a backbone, SSDLite with MobileNet and Faster R-CNN combined with Inception V2 and ResNet50. The results show that Faster R-CNN detects small object such as toy soldiers more successfully than SSD, and the training time of Faster R-CNN is much shorter than that of SSD.

Keywords: Object detectors · SSD · Faster R-CNN

1   Introduction

Convolutional neural network (CNN) [21] is a particular architecture of artificial neural networks, proposed by Yann LeCun in 1988. The key idea of CNN is that the local information in the image is important for understanding the content of the image so a filter is used when learning the model, focusing on the image, part by part, as a magnifying glass. The practical advantage of such approach is that CNN uses fewer parameters than fully-connected neural networks, which significantly improves learning time and reduces the amount of data needed to train the model.

Recently, after AlexNet [22] popularized deep neural networks by winning ImageNet competitions, convolutional neuronal networks have become the most popular model for image classification and object detection problems. Image classification predicts the existence of a class in a given image based on a model that is learned on a set of labeled images. There are several challenges associated with this task, including differences between objects of the same class, similarities between objects of different classes, object occlusions, different object sizes, various backgrounds. The appearance of an object on the image might change due to lighting conditions, position (height, angle) of the camera and distance from the camera and similar [19]. The detection of an object beside the prediction of the class to which the object belongs, provides information about its location in the image, so the challenge is to solve both the

© Springer Nature Switzerland AG 2019
S. Gievska and G. Madjarov (Eds.): ICT Innovations 2019, CCIS 1110, pp. 13–26, 2019.
https://doi.org/10.1007/978-3-030-33110-8_2

classification and location task. The detected object is most often labeled with the bounding box [23], but there are also detectors that segment objects at the pixel level and mark the object using its silhouette or shape [5, 14].

Some of today's most widely used deep convolution neural networks are Faster R-CNN, RFCN, SSD, Yolo, RetinaNet. These networks are unavoidable in tasks such as image classification [22] and object detection [26], analysis of sports scenes and activities of athletes [6], disease surveillance [25], surveillance and detection of suspicious behavior [19, 20], describing images [17], development and management of autonomous vehicles in robotics [10], and the like.

In this paper, we have focused on the problem of detecting small objects on footage taken by the camera of a mobile device or drones from a bird's eye view. These footages are today widely used when searching for missing persons in non-urban areas, border control, animal movement control, and the like.

In [1], drones were used to locate missing persons in search and rescue operations. Authors have used HOG descriptors [8]. In [3] the SPOT system is described. It uses an infrared camera mounted on an Unmanned Aerial Vehicle and Faster R-CNN to detect villains and control animals in images. A modified MobileNet architecture was used in [9] for body detection and localization in the sea. Images were shot both with an optical camera and a multi-spectral camera. In [33] YOLO was used for detection of objects on images taken from the air. In [24], three models of deep neural networks (SSD, Faster R-CNN, and RetinaNet) were analyzed for detection tasks on images collected by crewless aircraft. The authors showed that RetinaNet was faster and more accurate when detecting objects. The dependence analysis of Faster R-CNN, RFCN, and SSD speed and precision in case of running on different architectures was given in [18].

In this paper, we will approximate the problem of detecting small objects on bird-eye viewings or drone shots with the problem of detecting toy soldiers captured by the camera of a mobile device.

The rest of the paper is organized as follows: in Sect. 2. we will present the architecture of CNN networks, ResNet50, Inception and MobileNet with Faster R-CNN and SSD localization methods that are used in our research. We have examined their performance on a custom toy soldiers' dataset. The comparison of the detector performance and discussion are given in Sect. 3. The paper ends with a conclusion and the proposal for future research.

2   Convolutional Neural Networks

Convolutional Neural Networks (CNNs) are adapted to solve the problems of high-dimensional inputs and inputs that have many features such as in cases of image processing and object classification and detection. The CNN network consists of a convolution layer, after which the network has been named, activation and pooling layers, and at the end is most often one or more fully connected layers.

The convolution layer refers to a mathematical operator defined over two functions with real value arguments that give a modified version of one of the two original functions. The layer takes a map of the features (or input image) that convolves with a set of learned parameters resulting in a new two-dimensional map. A set of learned

parameters (weights and thresholds) are called filters or kernels. The filter is a 2D square matrix, small in size compared to the image to which it is applied (equal depths as well as the input). The filter consists of real values that represent the weights that need to be learned, such as a particular shape, color, edge in order to give the network good results.

The pooling layer is usually inserted between successive convolution layers, to reduce map resolution and increase spatial invariance - insensitivity to minor shifts (rotations, transformations) of features in the image as well as to reduce memory requirements for the implementation of the network. Along with the most commonly used methods (arithmetic mean and maximum [4]), there are several pooling methods used in CNN, such as Mixed Pooling, Lp Pooling, Stochastic Pooling, Spatial Pyramid Pooling and others [13].

The activation function propagates or stops the input value in a neuron depending on its shape. There is a broader range of neuron activation functions such as linear activation functions, jump functions, and sigmoidal functions. The jump functions and sigmoidal functions are a better choice for neural networks that perform classification while linear functions are often used in output layers where unlimited output is required. Newer architectures use activation functions behind each layer. One of the most commonly used activation functions in CNN is the ReLU (Rectified Linear Unit). In [13], the activation functions used in recent works are presented: Leaky Relu (LReLU), Parametric ReLU (PReLU), Randomized ReLU (RReLU), Exponential Linear Unit (ELU) and others.

A fully connected layer is the last layer in the network. The name of the fully connected layer indicates its configuration: all neurons in this layer are linked to all the outputs of the previous layer. Fully connected layers can be viewed as special types of convolution layers where all feature maps and all filters are 1×1.

Network hyperparameters are all parameters needed by the network and set before the network provides data for learning [2]. The hyper-parameters in convolution neural networks are learning rate, number of epochs, network layers, activation function, initialization weight, input pre-processing, pooling layers, error function.

Selecting the CNN network for feature extraction plays a vital role in object detection because the number of parameters and types of layers directly affect the memory, speed, and performance of the detector. In this paper, three types of network have been selected for feature extraction: ResNet50, Inception, and MobileNet.

2.1 ResNet

ResNet50 is a 50-layer Residual Network. There are other variants like ResNet101 and ResNet152 also [15]. The main innovation of ResNet is the skip connection. The skip connection in the Fig. 1 is labeled "identity." It allows the network to learn the identity function that allows passing the input through the block without passing through the other weight layers. This allows stacking additional layers and building a deeper network, as well as overcoming the vanishing gradient problem by allowing network to skip through layers if it feels they are less relevant in training. Vanishing gradients often occurs in deep networks if no adjustment is performed because during back-propagation gradient gets smaller and smaller and can make learning difficult.

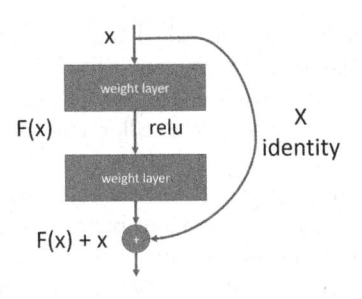

Fig. 1. A residual block, according to [15].

2.2   Inception

GoogLeNet has designed a module called Inception that approximates a sparse CNN with a normal dense construction, Fig. 2. The idea was to keep a small number of convolutional filters taking into account that only a small number of neurons are effective. The convolution filters of different sizes (5×5, 3×3, 1×1) were used to capture details on varied scales. In the versions Inception v2 and Inception v3, the authors have proposed several upgrades to increase the accuracy and reduce the computational complexity [28, 29].

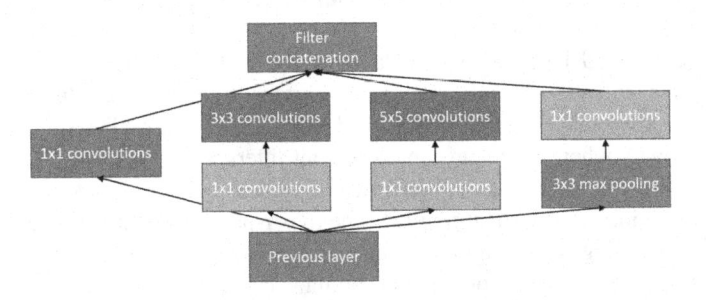

Fig. 2. Inception module, according to [28].

2.3   MobileNet

MobileNet is a lightweight architecture designed for use in a variety of mobile applications [16]. It filters the input channels by running a single convolution on each color channel instead of combining all three channels and flattening them all.

2.4   Faster R-CNN

In the earlier version of R-CNN [11] and Fast R-CNN [12], region proposals are generated by selective search (SS) [31] rather than using convolutional neural network (CNN). The SS algorithm was the "bottleneck" in region proposal process, so in the Faster R-CNN a separate convolution network (RPN) is used to propose regions. The RPN network then checks which location contains the object. Appropriate locations and bounding boxes are sent to the detection network that determines the class of

the object and returns the bounding box of that object. This kind of design has speed up the object detection.

2.5 Single Shot Detector

The Single Shot Detector (SSD) method for objects detection uses deep network that omits the stage of bounding box proposal and allows features extraction without losing accuracy. The approach assumes that potential objects can be located within the pre-defined bounding box of different size and side ratios centered in each location of feature map. The network for each bounding box determines the probability measure for the presence of each of the possible categories and adjusts the position of the box to frame the object better. In order to overcome the problems inherent in the difference in object sizes, the network makes decisions by combining prediction from several feature maps of different dimensions [23].

3 Comparison of SSD and Faster RCNN Detection Performance on Scenes of Toy Soldiers

We have tested and compared the accuracy of the object detector for a class of person (toy soldier) at different scene configurations, changing the number of object, their position, background complexity and lighting conditions. The goal is to select the appropriate model for future research on the detection of missing persons in rescue operations.

We used publicly available pre-trained models with corresponding weights learned on the Microsoft's common object and context (COCO) dataset [7] by transfer learning and fine-tune the model parameters on our data set.

We used the Tensorflow implementation [30] of the CNN model and the Python programming language in the Windows 10 × 64 environment. All models were trained on a laptop with the i5-7300HQ CPU and the Ge-Force GTX 1050Ti 4 GB GPU. The number of epochs and the training time differs among models and depends on loss. The parameters of each model have not been changed and were equal to the parameters of the original model.

3.1 Data Preprocessing

The data set contains 386 images shot by a mobile device camera (Samsung SM-G960F) at a 2160 × 2160 px resolution, without using a tripod. Each image contains multiple instances of toy soldiers, taken under different angles and different lighting conditions with a different background type from a uniform to complex (such as grassy surfaces). The images are divided into a learning and test set in a cutoff of 80:20, and their resolution is reduced to 720 × 720 px. In total, there are 798 toy soldiers in the images, of which 651 are in learning set and 147 in test.

The LabelImg tool was used to plot bounding box and create responsive XML files with stored xmin, xmax, ymin, ymax position for each layout. Images and corresponding XML files are then converted to TFRecord files that are implemented in the Tensorflow environment. TFRecord files merge all the images and notes into a single file, thus reducing the training time by eliminating the need of opening each file.

3.2   Methods

SSD with MobileNet
This method uses SSD for detection while the MobileNet network is used as a feature extractor. The output of MobileNets is processed using the SSD. We have tested the detection results of two versions of the MobileNet network (V1 and V2), referred to as ssd_mobilenet_v1 and ssd_mobilenet_v2. Both networks were pre-trained (ssd_mobilenet_v1_coco_2018_01_28 and ssd_mobilenet_v2_coco_2018_03_29) on COCO dataset of 1.5 million objects (80 categories) in 330,000 images. We trained the network using toy soldier's images width bounding box as input to the training algorithm. The network parameters include: prediction dropout probability 0.8, kernel size 1 and a box code size set to 4. The root mean square propagation optimization algorithm is used for optimizing the loss function with learning rate of 0.004 and decay factor 0.95. At the non-maximum suppression part of the network a score threshold of 1×10^{-8} is used with an intersection of union threshold of 0.6, both the classification and localization weights are set to 1. Ssd_mobilenet_v1 was trained for 17,105 steps and ssd_mobilenet_v2 for 10,123 steps.

SSD with Inception-V2
The combination of SSD and Inception-V2 is called SSD-Inception-V2. In this case, SSD is used for detection while Inception-V2 extracts features. We trained the network using predefined ssd_inception_v2_coco_2018_01_28 weights. The training process uses similar hyperparameters as SSD with MobileNet, except in this case of the kernel size that is set to 3. The network was trained for 6,437 steps.

SSDLite with MobileNet-V2
SSDLite [27] is a mobile version of the regular SSD, so all regular convolutions with detachable convolutions are replaced (depthwise followed by 1×1 projection) in SSD layers. This design is in line with the overall design of MobileNet and is considered to be much more efficient. Compared to SSD, it significantly reduces the number of parameters and computing costs. We trained the network using pre-trained ssdlite_mobilenet_v2_coco_2018_05_09 weights. Similar hyperparameters were used as before, and the network was trained for 14,223 steps.

Faster R-CNN with ResNet50
Faster R-CNN detection involves two phases. The first phase requires a region proposal network (RPN) that allows simultaneous prediction of object anchors and confidence (objectiveness) from some internal layers. For this purpose, a residual network with a depth of 50 layers (ResNet50) is used. The grid anchor size was 16×16 pixels with scales [0.25, 0.5, 1.0, 2.0], a non-maximum-suppression-IoU threshold was set to 0.7, the localization loss weight to 2.0, objectiveness weight to 1.0 with an initial crop size of 14, kernel size was 2 with strides set to 2. The second phase requires information from the first phase to predict the class label and the bounding box. We trained the network using pre-trained faster_rcnn_resnet50_coco_2018_01_28 weights. The IoU-threshold for prediction score was set to 0.6; the momentum (SGD) optimizer for optimizing the loss functions has initial learning rate set to 0.0002 and momentum value 0.9. The network was trained for 12,195 steps.

Faster R-CNN with Inception-V2

Faster R-CNN uses the Inception V2 feature extractor to get features from the input image. The middle layer of the Inception module uses the RPN network component to predict the object anchor and confidences. As in previous cases, the network was trained with pre-trained fast-er_rcnn_inception_v2_coco_2018_01_28 weights. Similar hyperparameters were used as in case of Faster R-CNN with ResNet50 and the learning process lasted for 33,366 steps.

3.3   Results and Discussion

We compared the results of the SSD model and the Faster RCNN object detector based on CNNs on our toy soldiers test set concerning mean average precision (mAP) [32]. A detection is considered as true positive when more than half of the area belonging to the soldier is inside the detected bounding box. Detectors performance are also evaluated in terms of recall, precision and F1 score.

$$F1 = \frac{2 \cdot \text{Recall} \cdot \text{Precision}}{(\text{Recall} + \text{Precision})} \tag{1}$$

Figure 3 shows a comparison of results of models that were additionally learned on our learning set with original models trained on the COCO dataset. The results show a significant increase in the average precision of all models after training on our dataset. The best results of over 96% were achieved with the faster_rcnn network. The implementation of faster_rcnn with Resnet50 proved to be somewhat successful than architecture with the Inception Network. Faster_rcnn has also shown the best classification results concerning F1 score and Recall [19], Fig. 4. All classification result of all models in terms of precision, recall and F1 score are shown in Fig. 4.

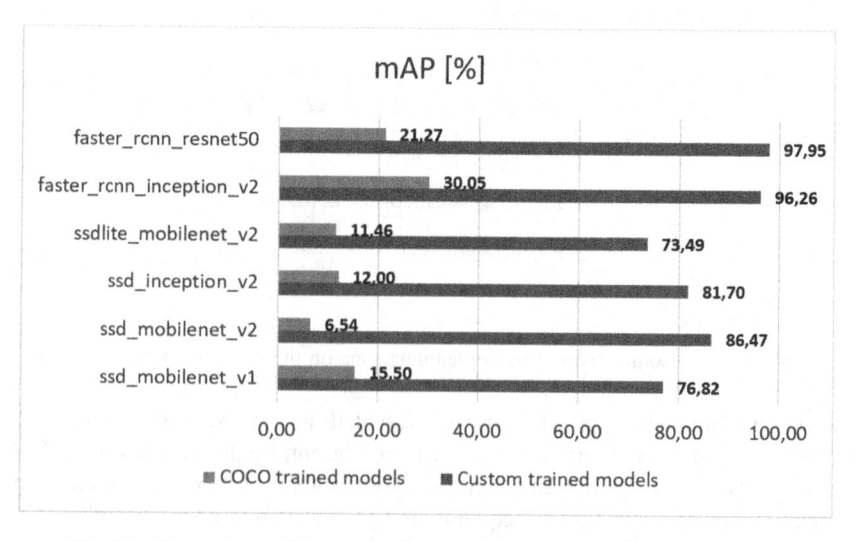

Fig. 3. Comparison of the evaluation result of the toy soldier's detection.

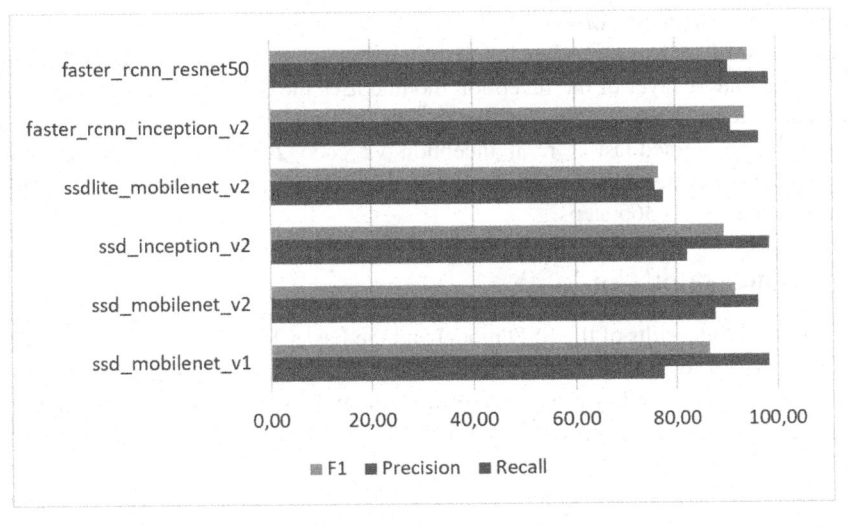

Fig. 4. Comparison of the results of the trained model detection concerning the F1, Precision and Recall metrics.

Figure 5 shows the time required to train the model on our learning set. The least amount of time was needed to learn the faster_rcnn model. The longest, more than 3.5 times longer than learning the fast_rcnn model, was needed to learn the ssdlitle-mobilenet_v2 model.

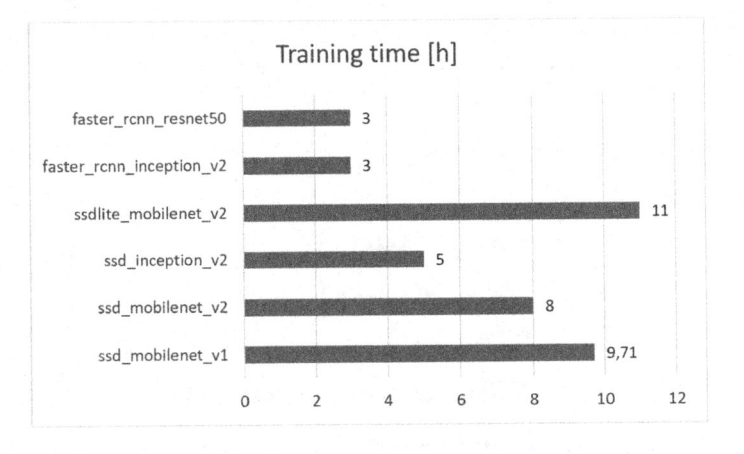

Fig. 5. Comparison of model learning time on the custom dataset.

Model performances are additionally presented in two scenarios: simple and complex. The simple scenario has a uniform background color and up to 8 visible objects near the camera, which may overlap. A complex scenario is considered when the number of objects in a scene is equal to and greater than 9, away from the camera and with occlusions. A Fig. 6 shows an example of the detection results in the case of a

simple scenario. The images marked A through F show the same scenarios with a uniform background with a wooden pattern and five soldiers in different poses such as walking with a gun, shooting, crawling and lying down.

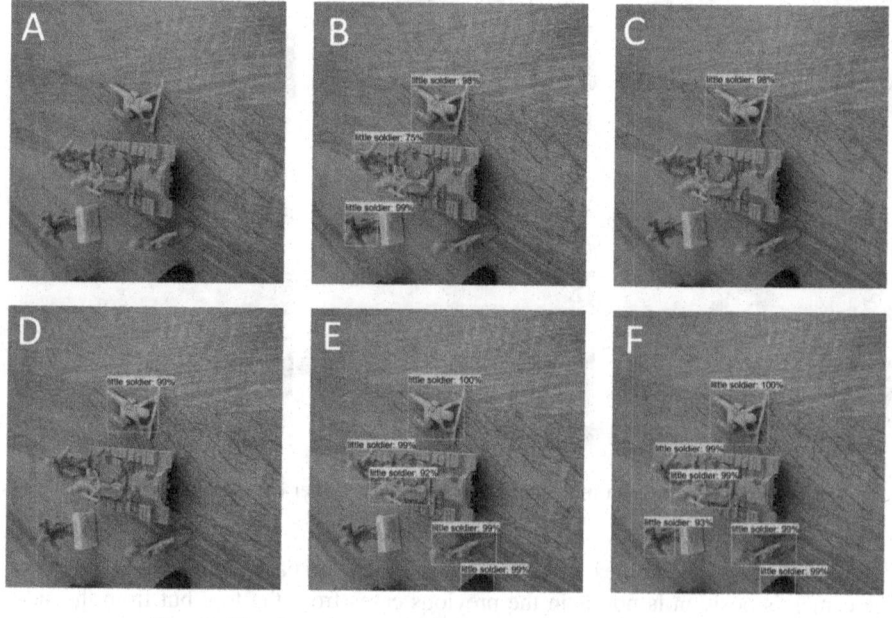

Fig. 6. The detection results in the case of a simple scenario.

In all the images, the results of individual models are indicated in the following way:

- A – ssd_mobilenet_v1,
- B – ssd_mobilenet_v2,
- C – ssdlite_mobilenet_v2,
- D – ssd_inception_v2,
- E – faster_rcnn_resnet_50,
- F – faster_rcnn_inception_v2.

In figure A, no soldier was detected, B has 3 of 5 true positive (TP) detections, C and D only one detected soldier, while E has 4 TP with one false positive (FP) detection, and F all positive detections with one FP.

In the case of a uniform background with higher contrast to soldiers, as in Fig. 7 all models have detected with greater success in comparison to the previous case, even though in this example, we have a higher number of soldiers and at a greater distance. There were 11 soldiers on the scene, but no model detected a soldier on a tank of the identical color. The best results were achieved in E and F images with 10 successful detections, then model B with 7, and then models A and C follow. The sequence of the success of the model is similar to the one in the previous example.

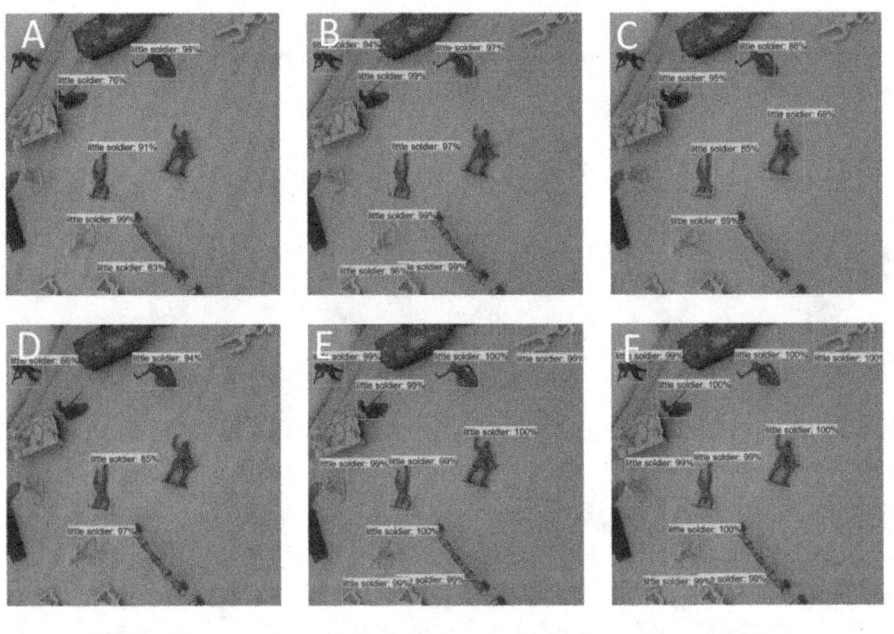

Fig. 7. The case of a uniform background with higher contrast to soldiers.

Figure 8 shows a complex scene in which soldiers are partially covered with grass. The camera's position is not as in the previous cases from the top, but from the side.

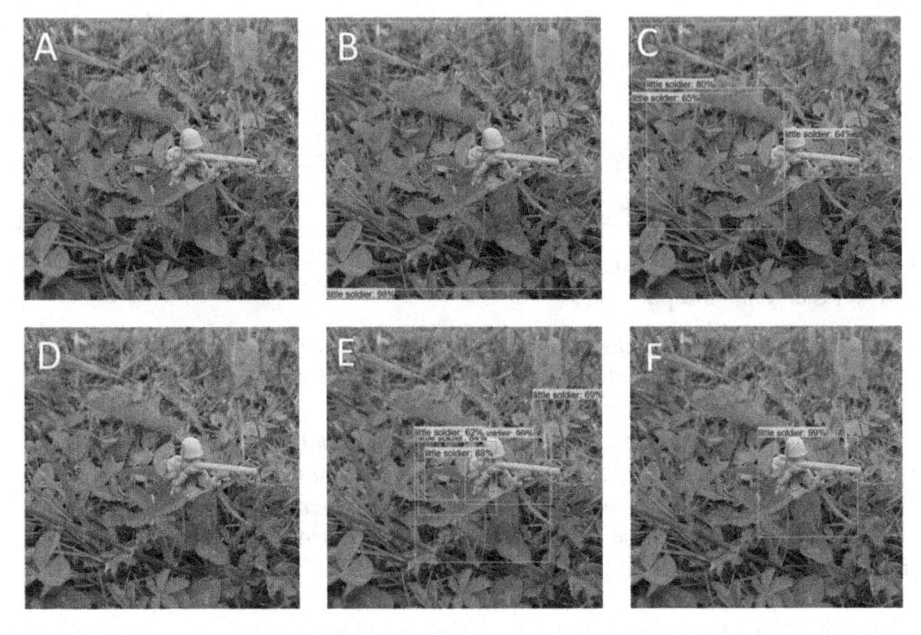

Fig. 8. A complex scene in which soldiers are partially covered with grass.

The models in Figures A and D did not have any detection, while B detected almost the entire image as a soldier. C has one positive and two false detections, E has repeatedly detected the same object, but with different rectangle size and has a false detection, and F has an accurate, true positive detection.

Figure 9 shows two scenes with a camera positioned from a bird's perspective at a greater distance than in earlier cases and with 8 soldiers on a uniform blue background. Model A detected only one soldier (Fig. 9a) whereas B, C, and D had no detection (Fig. 9b). Model E detected all soldiers (Fig. 9d), while F detects all soldiers plus three false detections (Fig. 9e).

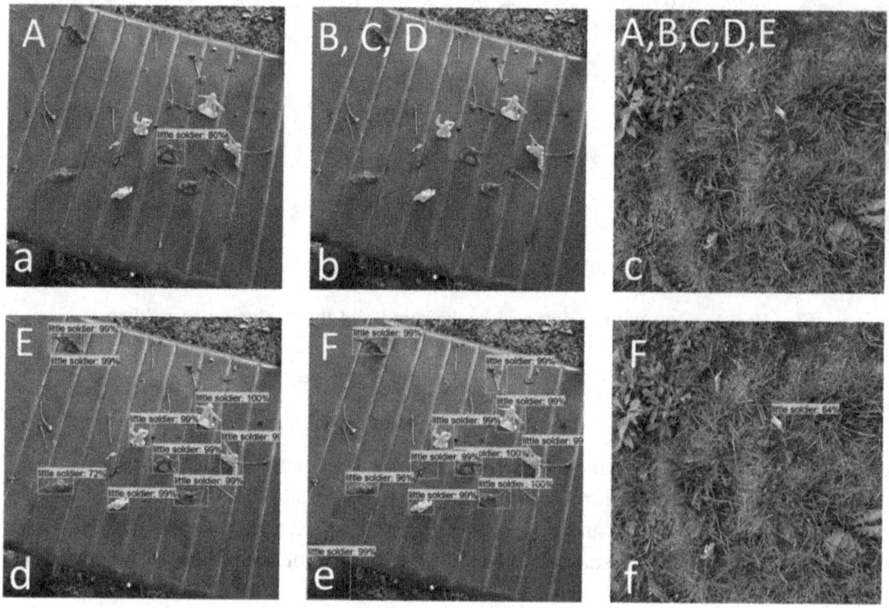

Fig. 9. Two scenes with a camera positioned from a bird's perspective.

In the second scene recorded from a greater distance on the grass, Fig. 9c and f, only F detects one soldier out of 7 possible. Examples show that all models have problems with object detection when an object is less than 50 px in height or width, especially when the contrast of the subject and background is not significant, and when the background is more complex than in the case of grass.

Figure 10 is an example of a scene with two soldiers with a cluttered background. E and F models detected both soldiers with a probability of detection of 100% (Fig. 10a, c and e), while model B detected only one soldier (Fig. 10b and d) and other models failed to detect anything. Figure 10e shows a higher contrast between the soldiers and the background, but this did not help models A, B, C, D to have a successful detection. Figure 10f shows the occlusion of soldiers; however, models B, D, E, and F were able to detect them.

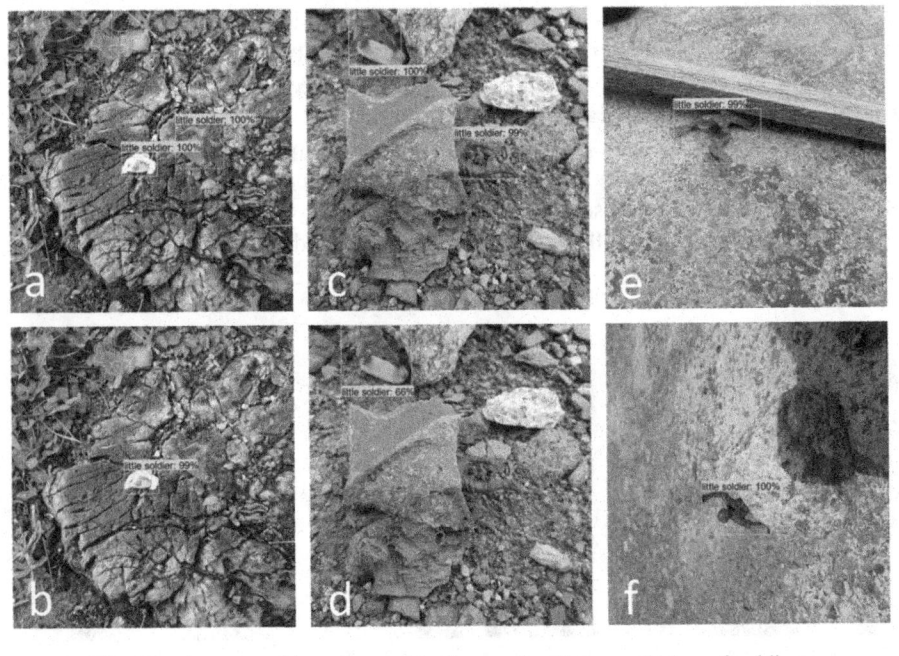

Fig. 10. A scene with a cluttered background and the occlusion of soldiers.

4   Conclusion

Recordings taken from the air today are used mainly in search of missing persons, in mountain rescue, in the control of animal movement, and the like. The ability to automatically detect persons and objects on the images taken from a bird's perspective would greatly facilitate the search and rescue of people or the control of people and animals.

CNN networks have proven successful in classification and object detection tasks on general-purpose images, and in this paper, we have tested their performance in detecting toy soldiers taken from the bird's eye view. On the custom dataset, we compared the performance of ResNet50, Inception, and MobileNet networks with Faster RCNN and SSD methods of localization. The analysis of the obtained results shows that Faster RCNN is more suitable for detection because it detects toy soldiers more successfully. The configuration with the Inception network is more successful than the configuration with ResNet50. The problem with this method is that it requires more time and computation power.

The examples also show the background effect on detection accuracy. With a uniform background and higher color contrast, detection of all models is significantly successful than in case of detection at a greater distance, on the grass, and with semi-hidden objects. In future work, we will try to find a way to solve this problem.

This paper provides a promising base ground for further research in real-time detection of missing persons in search and rescue operations. We plan to investigate the further use of different detection methods (speed, accuracy) on the android system.

Acknowledgment. This research was supported by Croatian Science Foundation under the project IP-2016-06-8345 "Automatic recognition of actions and activities in multimedia content from the sports domain" (RAASS) and by the University of Rijeka under the project number 18-222-1385.

References

1. Andriluka, M., et al.: Vision based victim detection from unmanned aerial vehicles. In: 2010 IEEE/RSJ International Conference on Intelligent Robots and Systems, pp. 1740–1747. IEEE (2010)
2. Bengio, Y.: Practical recommendations for gradient-based training of deep architectures. In: Montavon, G., Orr, Geneviève B., Müller, K.-R. (eds.) Neural Networks: Tricks of the Trade. LNCS, vol. 7700, pp. 437–478. Springer, Heidelberg (2012). https://doi.org/10.1007/978-3-642-35289-8_26
3. Bondi, E., et al.: Spot poachers in action: augmenting conservation drones with automatic detection in near real time. In: Thirty-Second AAAI Conference on Artificial Intelligence (2018)
4. Boureau, Y., Ponce, J., LeCun, Y.: A theoretical analysis of feature pooling in vision algorithms. In: Proceedings of International Conference on Machine learning (ICML10), vol. 10 (2010)
5. Burić, M., Pobar, M., Ivašić-Kos, M.: Ball detection using YOLO and mask R-CNN. In: 2018 International Conference on Computational Science and Computational Intelligence (CSCI 2018), pp. 319–323 (2018)
6. Burić, M., Pobar, M., Ivašić-Kos, M.: Adapting YOLO network for ball and player detection. In: Proceedings of the 8th International Conference on Pattern Recognition Applications and Methods (ICPRAM 2019), pp. 845–851. SciTePress, Portugal (2019)
7. Cocodataset. http://cocodataset.org/
8. Dalal, N., Triggs, B.: Histograms of oriented gradients for human detection. In: International Conference on Computer Vision & Pattern Recognition, pp. 886–893 (2005)
9. Gallego, A.J., Pertusa, A., Gil, P., Fisher, R.B.: Detection of bodies in maritime rescue operations using unmanned aerial vehicles with multispectral cameras. J. Field Robot. **36**(4), 782–796 (2019)
10. Gao, H., Cheng, B., Wang, J., Li, K., Zhao, J., Li, D.: Object classification using CNN-based fusion of vision and LIDAR in autonomous vehicle environment. IEEE Trans. Ind. Inform. **14**(9), 4224–4231 (2018)
11. Girshick, R., Donahue, J., Darrell, T., Malik, J.: Rich feature hierarchies for accurate object detection and semantic segmentation. In: Proceedings of the IEEE Conference on Computer Vision and Pattern Recognition. pp. 580–587 (2014)
12. Girshick, R.: Fast R-CNN. In: Proceedings of the IEEE International Conference on Computer Vision, pp. 1440–1448 (2015)
13. Gu, J., et al.: Recent advances in convolutional neural networks. Pattern Recogn. **77**, 354–377 (2018)
14. He, K., Gkioxari, G., Dollar, P., Girshick, R.: Mask R-CNN. In: Proceedings of the IEEE International Conference on Computer Vision, pp. 2961–2969 (2017)
15. He, K., Zhang, X., Ren, S., Sun, J.: Deep residual learning for image recognition. In: Proceedings of the IEEE Conference on Computer Vision and Pattern Recognition, pp. 770–778 (2016)

16. Howard, A.G., et al.: MobileNets: efficient convolutional neural networks for mobile vision applications. arXiv preprint arXiv:1704.04861 (2017)
17. Hrga, I., Ivašić-Kos, M.: Deep mage captioning: an overview. In: 42nd International ICT Convention–MIPRO 2019-CIS-Intelligent Systems (2019)
18. Huang, J., et al.: Speed/accuracy trade-offs for modern convolutional object detectors. In: Proceedings of the IEEE Conference on Computer Vision and Pattern Recognition, pp. 7310–7311 (2017)
19. Ivašić-Kos, M., Ipšić, I., Ribarić, S.: A knowledge-based multi-layered image annotation system. Expert Syst. Appl. **42**(24), 9539–9553 (2015)
20. Ivašic-Kos, M., Krišto, M., Pobar, M.: Human detection in thermal imaging using YOLO. In: Proceedings of the 2019 5th International Conference on Computer and Technology Applications, pp. 20–24. ACM (2019)
21. Johnson, J., Karpathy, A.: Convolutional Neural Networks, Stanford Computer Science. https://cs231n.github.io/convolutional-networks
22. Krizhevsky, A., Sutskever, I., Hinton, G.E.: ImageNet classification with deep convolutional neural networks. In: Advances in Neural Information Processing Systems, pp. 1097–1105 (2012)
23. Liu, W., et al.: SSD: single shot multibox detector. In: Leibe, B., Matas, J., Sebe, N., Welling, M. (eds.) ECCV 2016. LNCS, vol. 9905, pp. 21–37. Springer, Cham (2016). https://doi.org/10.1007/978-3-319-46448-0_2
24. Radovic, M., Adarkwa, O., Wang, Q.: Object recognition in aerial images using convolutional neural networks. J. Imaging **3**(2), 21 (2017)
25. Ramcharan, A., et al.: Assessing a mobile-based deep learning model for plant disease surveillance (2018)
26. Ren, S., He, K., Girshick, R., Sun, J.: Faster R-CNN: towards real-time object detection with region proposal networks. In: Advances in Neural Information Processing Systems, pp. 91–99 (2015)
27. Sandler, M., Howard, A., Zhu, M., Zhmoginov, A., Chen, L.C.: MobileNetV2: inverted residuals and linear bottlenecks. In Proceedings of the IEEE Conference on Computer Vision and Pattern Recognition, pp. 4510–4520 (2018)
28. Szegedy, C., et al.: Going deeper with convolutions. In: Proceedings of the IEEE Conference on Computer Vision and Pattern Recognition, pp. 1–9 (2015)
29. Szegedy, C., Vanhoucke, V., Ioffe, S., Shlens, J., Wojna, Z.: Rethinking the inception architecture for computer vision. In: Proceedings of the IEEE Conference on Computer Vision and Pattern Recognition, pp. 2818–2826 (2016)
30. Tensorflow object detection models zoo. https://github.com/tensorflow/models/blob/master/research/object_detection/g3doc/detection_model_zoo.md
31. Uijlings, J.R., Van De Sande, K.E., Gevers, T., Smeulders, A.W.: Selective search for object recognition. Int. J. Comput. Vis. **104**(2), 154–171 (2013)
32. Visual Object Classes Challenge 2012 (VOC2012). http://host.robots.ox.ac.uk/pascal/VOC/voc2012/
33. Wang, X., Cheng, P., Liu, X., Uzochukwu, B.: Fast and accurate, convolutional neural network based approach for object detection from UAV. In: IECON 2018-44th Annual Conference of the IEEE Industrial Electronics Society. pp. 3171–3175. IEEE (2018)

Prediction of Student Success Through Analysis of Moodle Logs: Case Study

Neslihan Ademi[1]([✉]), Suzana Loshkovska[2],
and Slobodan Kalajdziski[2]

[1] International Balkan University, Skopje, North Macedonia
neslihan@ibu.edu.mk
[2] Ss. Cyril and Methodius University, Skopje, North Macedonia
{suzana.loshkovska,slobodan.kalajdziski}@finki.ukim.mk

Abstract. Data mining together with learning analytics are emerging topics because of the huge amount of educational data coming from learning management systems. This paper presents a case study about students' grade prediction by using data mining methods. Data obtained from Moodle log files are explored to understand the trends and effects of students' activities on Moodle learning management system. Correlations of system activities with the student success are found. Data is classified and modeled by using decision tree, Bayesian Network and Support Vector Machine algorithms. After training the model with a one-year course activity data, next years' grades are predicted. We found that Decision tree classification gives the best accuracy on the test data for the prediction.

Keywords: Data Mining · Learning Analytics · LMS · E-learning · Prediction

1 Introduction

With the rapid development of e-technologies distance learning has gained a bigger role in the education. E-learning systems are used even in the formal education settings very often as they offer instructor great variety of opportunities. Learning management systems (LMSs) are one of the most popular methods as they offer the opportunities to facilitate to distribute information to students, communicate among participants in a course, produce content material, prepare assignments and tests, engage in discussions, manage distance classes and enable collaborative learning with forums, chats, file storage areas. One of the most commonly used LMS is Moodle (modular object-oriented developmental learning environment), which is a free and open learning management system.

E-learning systems like Moodle contain a huge amount of data related to the students. This data gives a big opportunity to analyze students' behavior and understand the characteristics and behavior of students.

This huge amount of data, in other words Big Data in education emerged two research areas: Educational Data Mining (EDM) and Learning Analytics (LA). EDM is concerned with developing tools for the discovery of patterns in educational data [15]. LA is concerned with the measurement, collection and analysis and reporting of data

© Springer Nature Switzerland AG 2019
S. Gievska and G. Madjarov (Eds.): ICT Innovations 2019, CCIS 1110, pp. 27–40, 2019.
https://doi.org/10.1007/978-3-030-33110-8_3

about learners to understand, to optimize learning and the learning environment [9]. Learning analytics can be used to understand students' behaviors in learning management systems.

Especially log data in Moodle can be used as a basis for creating educational data which gives us the information about students' online behavior. In all LMSs, every student posses' digital profile, and every student can access the LMS by using their personal account. All activities performed by the students are saved in log files. Collected log data provide a descriptive overview of human behavior. Simply observing behavior at scale provides insights about how people interact with existing systems and services [4]. Generally observational log studies contain partitioning log data; by time and by user. Partitioning by time helps us to understand significant temporal features, such as periodicities (including consistent daily, weekly, and yearly patterns) and sharp changes in behavior during important events. It gives an up-to-the-minute picture of how people are behaving with a system from log data by comparing past and current behavior. It is also interesting to partition log data by user characteristics [4].

Recently there are many studies in the literature about log analysis in e-learning environments. In [14], authors explain educational data mining with a case study on Moodle logs for visualization, clustering and classification purposes. Another study about log analysis on e-learning systems demonstrates the disengagement of the students by looking their online behavior [3]. In [11], authors use learning analytics for monitoring students' online participation from Moodle logs and discover frequent navigational patterns by sequential pattern mining techniques. In study [2], clusters are used for profiling the students according to their learning styles as deep learners and surface learners with the help of Moodle log data. The study [7] aims to detect the relationship between the observed variables and students' final grades and to find out the impact of a particular activity in an LMS on the final grade considering also other data such as gender. In [10], authors used Excel macros to analyze and visualize Moodle activities based on metrics such as total page views, unique users, unique actions, IP addresses, unique pages, average session length and bounce rate. The study [8] investigates the impact of students' exploration strategies on learning and proposes a probabilistic model jointly representing student knowledge and strategies based on data collected from an interactive computer-based game. The study [13] analyzes log data obtained from various courses of an university using Moodle platform together with the demographic profiles of students and compares them with their activity level in order to find how these attributes affect students' level of activity. In [12], authors suggest an analytics package which can be integrated with Moodle, to fetch the log files produced from Moodle continuously and produce the results of learning trends of the students. Some studies like in [5], use log data for the grade prediction by using decision trees.

In our study we focus on the engagement of the students by evaluating number of their activities and we make grade prediction upon these activities. The purpose of this study is to predict students' success in terms of grade and also to figure out the students who are about to drop out. Differently from the above mentioned studies, in this paper we use one years' data to predict the next year. The paper presents overall steps of the analysis of Moodle log data by using several data mining methods and tools as a case study. The second section defines the used methodology for the analysis, while the third section gives results and discussion, finally the last section is the conclusion of the paper.

2 Methodology

Generally e-learning data mining process consists of four steps like in the general data mining process: (1) collect data, (2) pre-process the data, (3) apply data mining, (4) interpret, evaluate and deploy the results [14]. When the log files are considered for data mining pre-processing part is more complicated because of the structure of the log files. Log files contain one row for each activity in the system, so for the user based analysis logs should be filtered and the data should be transformed into a data frame based on user and number of each activity of the user. In Fig. 1 we give the methodology describing the work flow of the steps used for the case study.

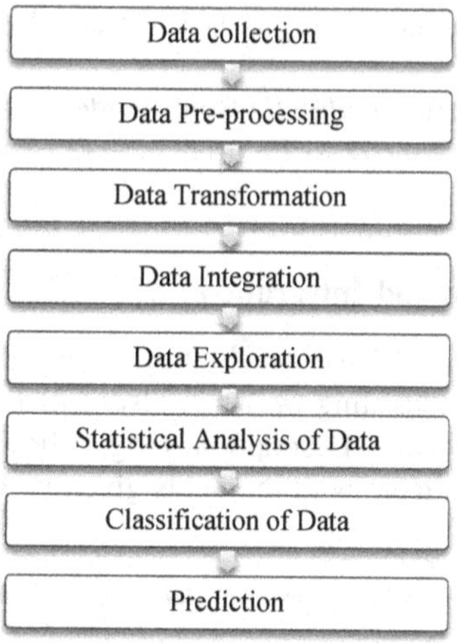

Fig. 1. Flow chart of the used methodology.

2.1 Data Collection

For the study, two log files are taken from Moodle which is installed and used at the Faculty of Computer Science and Engineering at the University of Ss. Cyril and Methodius in Skopje. Log files were extracted in .csv format and they contained all activities of the students from a one semester bachelor's degree course of User Interfaces at two academic years; 2016–2017 and 2017–2018. Teaching process was designed as blended learning. Moodle was used to support classroom teaching to distribute course material, lectures, homework, laboratory exercises and to provide discussion through the forums. A total of 260 students registered with Moodle for the course of User Interfaces for the first year and 206 students for the second year.

The standard retrieved fields in the log files are: Time, User full name, Affected user, Event context, Component, Event name, Description, Origin, IP address. The retrieved data for the academic year 2016–2017 was composed of 161.007 rows, for the academic year 2017–2018 was composed of 165.000 rows each with a filled value in every of the above-mentioned column fields.

Separately, another file is used which contained scores and grades of the students from the course for both academic years.

2.2 Data Pre-processing

Several pre-processing steps are applied to the log files, to keep on relevant and correct information. The first step was to convert all Cyrillic letters found in the original log files into Latin letters, as they were not recognized by R packages. The second step was to remove the actions logged by instructors and administrators selectively by filtering them using *sqldf* package [6], as we want to analyze only the students' actions. Log data produced by the system is also removed by filtering the data where component field is system. Required fields are extracted and duplicate records are removed. The number of remaining rows after filtering is 144.522 and 146.711 for the first and the second file respectively.

As the last step, *userID* and *moduleID* from the description field is filtered so we generated new columns for *userID* to be used instead of user full name to provide anonymity of the data.

2.3 Data Transformation and Integration

For data transformation *sqldf* package which allows complex database queries is used in RStudio. Raw data was consisting of Time, User full name, Affected user, Event context, Component, Event name, Description, Origin, IP address. After filtering and transformation, we created a table with the fields given in Table 1.

Table 1. Attributes after data transformation.

Name	Description
UserID	ID number of the student
Visits	Total number of visits by the student
Quizzes	Number of quizzes taken by the student
Assignments	Number of submitted assignments by the student
ForumCreated	Number of forum creations by the student
ForumView	Number of forum views by the student
CourseView	Number of course views by the student
FileSubmission	Number of file submissions by the student
GradeView	Number of grade views by the student

Transformed data is integrated with the data containing course grades. Table 1 is updated by adding one more attribute "Grade". Grades are in the range of 5–10 where 5 represent failure of the student, while 10 is the highest score.

2.4   Data Exploration and Statistical Analysis

The results of this step are obtained by using the *sqldf* package and the lattice package in R Studio and also with WEKA software.

In a previous study [1], we explored 2016–2017 data in details by visualizing views of lectures for each week, views of lab exercises, total visit frequency, distribution of the grades, quiz, assignment, forum, file submission, grade view frequencies and distribution of weekly visits. Here we just give the summary statistics of both years' data and correlation matrices of visits, quizzes, assignments, forum creations, forum views, file submissions, grade views with the course grade as they show us the similarity of the activities and correlations of both years. Also, we tried to use those summary statistics (5-point summary of data) for classification when labeling the classes. Standard discretization/binning methods in WEKA could not be used, because the data of the next year has different summary statistics. This is why 5 point summary of the data is important for both of the years' data.

Correlations are found by using Pearson correlation test. Equation 1 gives the formula for Pearson correlation coefficient which calculates the correlation between two variables X and Y. The value of the Pearson coefficient is always between −1 and +1, where $r = -1$ or $r = +1$ indicates a perfect linear relationship, where sign indicates the direction, while $r = 0$ indicates no linear relationship.

$$r = \frac{\sum (x - \bar{X})(y - \bar{Y})}{\sqrt{\left[\sum (x - \bar{X})^2\right]\left[\sum (y - \bar{Y})^2\right]}} \tag{1}$$

2.5   Classification of Data

Classification is a supervised learning method which uses labels to group the data. The attributes of both years shown in Table 1 are labeled in terms of LOW, MEDIUM, HIGH; or in terms of YES and NO. For the labeling process 5 point summary statistics of the data is used. The attributes Visits, Assignments, Course views and Forum views are labeled as LOW if the value is between min to Q1, MEDIUM if the value is in between Q1-Q3 and HIGH if the value is in between Q3-max, where Q1 represents the first quartile and Q3 represents the third quartile of the data attributes in 5-point summary statistics. The attributes Forum created and Grade view are labeled in terms of YES or NO as their number is very low and their first quartile values are 0. So, all activities of the students are labeled with in their study year.

The grade attribute in the first data file are left in form of numbers 5, 6, 7, 8, 9 and 10 just changed from numeric to nominal by using WEKA's NumericToNominal function so that the text values of the grades are considered. The grade attribute is set as the target attribute to be predicted.

We also tried the model with the labeling of the grades in terms of FAIL, GOOD and EXCELLENT, and in terms of PASS and FAIL to compare the accuracy of the model in different number of labels.

2.6 Prediction

The data from 2016–2017 is used as training data to generate a model and this model is used to predict the grades of the next year. In this step WEKA is used. For the classification, J48 Decision Tree (DT), Bayesian Network (BN) and Support Vector Machine (SVM) algorithms are applied. In all cases the model is trained with 10 fold cross-validation. After training the first data file with above mentioned classification algorithms, it is saved as a model in WEKA; so this model can be loaded and used any time on a test data set. The second file for 2017–2018 data also contained grade attribute. The content of this attribute is changed by a "?" and this file is used as a test set. So, the created model from the first-year data is used to predict the grades of the second year by using DT, BN and SVM. In the end the grades generated by the model and the real grades are compared.

2.7 Evaluation of the Predictions

Evaluation of the classification models are done in terms of True Positive (TP) rate, False Positive (FP) rate, Precision, Recall and F-Measure parameters for the algorithms DT, BN and SVM. After the models are created and applied on the test data in WEKA, results are given in form of confusion matrices for the predicted data vs. real data. Confusion matrices contain Recall and Precision parameters. In the end three classification models are compared in terms of overall accuracy and Kappa Index values.

3 Results and Discussion

In the complete process we used two software; RStudio and WEKA which are both open source under GNU. RStudio is a development interface for R language which allows easy manipulation of data. WEKA is useful in terms of its ready interface, which contains machine learning algorithms for data mining tasks.

3.1 Exploratory Data Analysis of Student Activities

Tables 2 and 3 gives the summary statistics of data in years 2016–2017 and 2017–2018 respectively. Summary statistics contain 5-point summary of the data which is minimum value, first quartile, mean, third quartile and maximum value; and additionally median value.

Table 2. Summary statistics of 2016–2017 data.

	Visits	Quizzes	Assignment	Forum created	Forum view	Course view	File submission	Grade view	Grade
Min	1	0	0	0	0	0	0	0	5
Q1	206.8	1	7	0	3	45.75	7	0	5
Median	281.5	2	11	0	7	65.50	11	1	6
Mean	301.9	1.938	10.87	0.1769	11.7	71.8	10.87	2.85	6.77
Q3	390.2	3	15	0	14	93.50	15.00	3	8
Max	1188	5	25	5	77	256	25.00	74	10

Table 3. Summary statistics of 2017–2018 data.

	Visits	Quizzes	Assignment	Forum created	Forum view	Course view	File submission	Grade view	Grade
Min	1.0	0	0	0	0	0	0	0	5
Q1	258.2	7	7	0	2	66.25	7	0	6
Median	362.5	11	12	0	5	95.00	12	1	4
Mean	402.5	9.53	11.35	0.01	8.28	106.03	11.35	3.64	7.3
Q3	494.8	13	16	0	11	127.75	16	4	9
Max	2853	15	31	1	63	531.00	31	52	10

Figure 2 shows the activities of the students in terms of grades for the year 2016–2017. To generate this graph grades are grouped in terms of PASS and FAIL. The blue color represents the failed students while the red color represents passed students. Figure 3 shows the data when grades are classified in terms of numbers from 5 to 10.

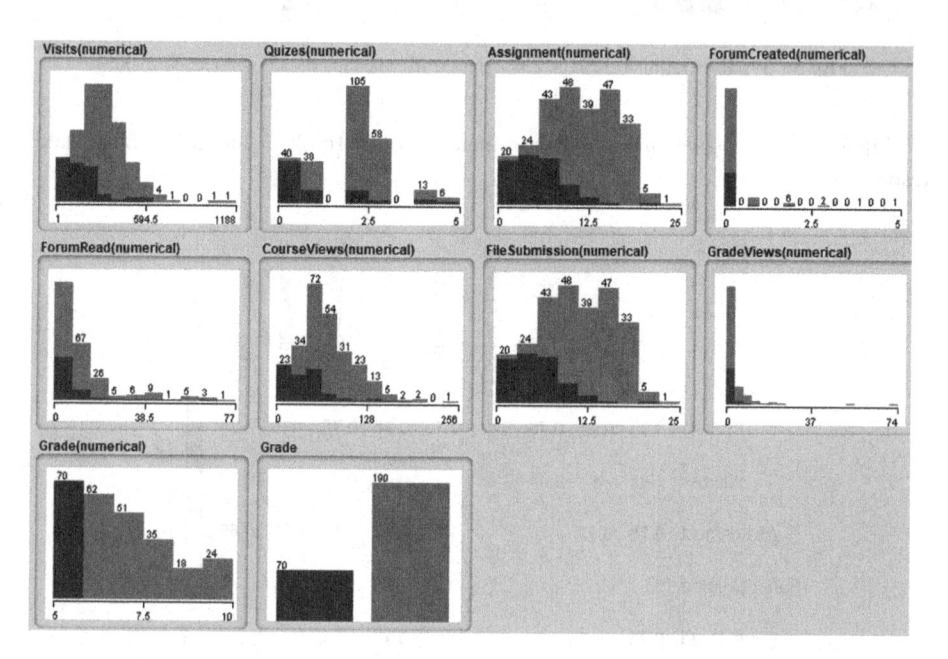

Fig. 2. WEKA output of 2016–2017 data with two classes (Pass/Fail).

Visits
169
28
63
Quizes
163
78
19
Assignment
104
104
62
ForumCreated
29
231
ForumRead
233
9
18
CourseViews
180
71
9
FileSubmission
104
104
62
GradeViews
246
4
11
Grade
70
62
51
35
18
24

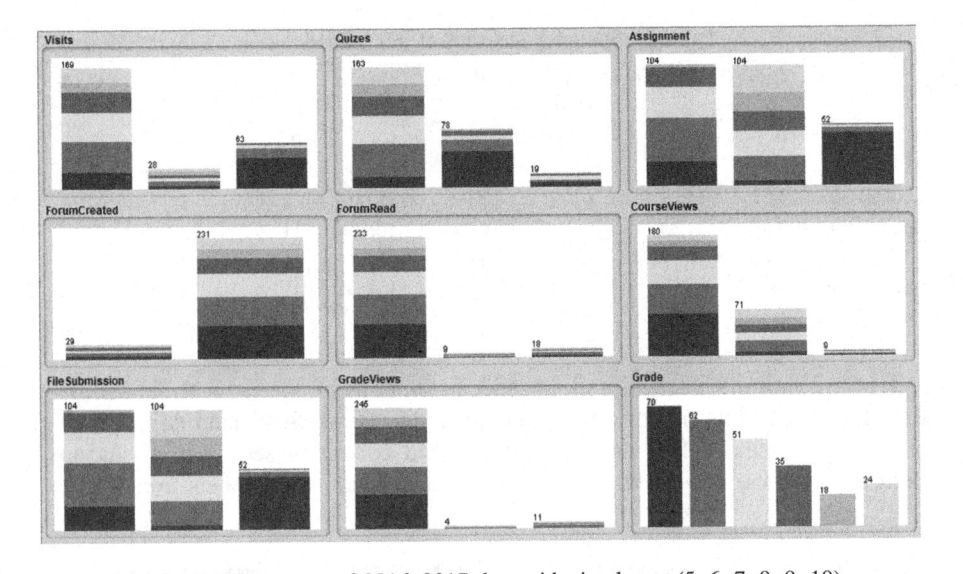

Fig. 3. WEKA output of 2016–2017 data with six classes (5, 6, 7, 8, 9, 10).

Figures 4 and 5 gives the correlation matrices of 2016–2017 and 2017–2018 data respectively.

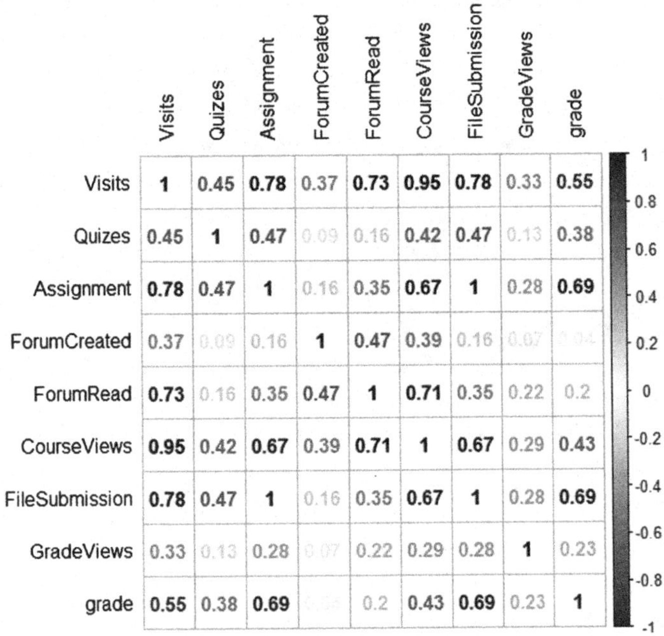

Visits Quizes Assignment ForumCreated ForumRead CourseViews FileSubmission GradeViews grade
Visits 1 0.45 0.78 0.37 0.73 0.95 0.78 0.33 0.55
Quizes 0.45 1 0.47 0.09 0.16 0.42 0.47 0.13 0.38
Assignment 0.78 0.47 1 0.16 0.35 0.67 1 0.28 0.69
ForumCreated 0.37 0.09 0.16 1 0.47 0.39 0.16 0.07 0.04
ForumRead 0.73 0.16 0.35 0.47 1 0.71 0.35 0.22 0.2
CourseViews 0.95 0.42 0.67 0.39 0.71 1 0.67 0.29 0.43
FileSubmission 0.78 0.47 1 0.16 0.35 0.67 1 0.28 0.69
GradeViews 0.33 0.13 0.28 0.07 0.22 0.29 0.28 1 0.23
grade 0.55 0.38 0.69 0.2 0.43 0.69 0.23 1
1
0.8
0.6
0.4
0.2
0
-0.2
-0.4
-0.6
-0.8
-1

Fig. 4. Correlation analysis of 2016–2017 data.

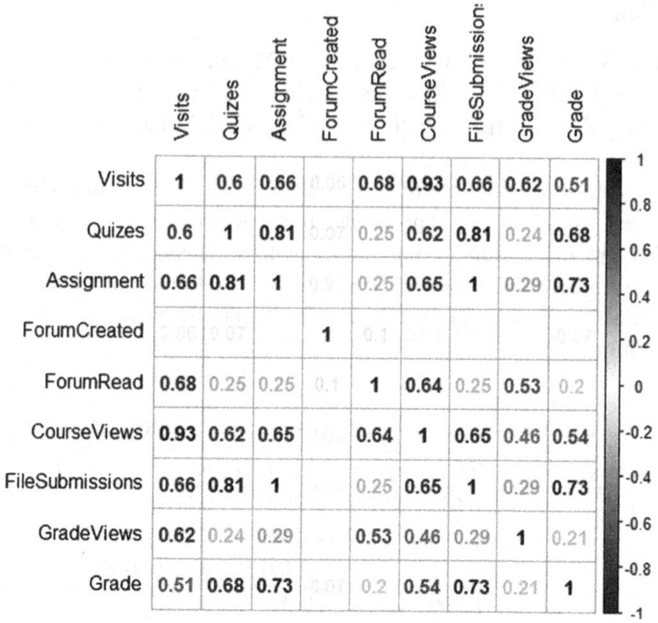

	Visits	Quizes	Assignment	ForumCreated	ForumRead	CourseViews	FileSubmission:	GradeViews	Grade
Visits	1	0.6	0.66		0.68	0.93	0.66	0.62	0.51
Quizes	0.6	1	0.81	0.07	0.25	0.62	0.81	0.24	0.68
Assignment	0.66	0.81	1		0.25	0.65	1	0.29	0.73
ForumCreated		0.07		1	0.1				
ForumRead	0.68	0.25	0.25	0.1	1	0.64	0.25	0.53	0.2
CourseViews	0.93	0.62	0.65		0.64	1	0.65	0.46	0.54
FileSubmissions	0.66	0.81	1		0.25	0.65	1	0.29	0.73
GradeViews	0.62	0.24	0.29		0.53	0.46	0.29	1	0.21
Grade	0.51	0.68	0.73	0.07	0.2	0.54	0.73	0.21	1

Fig. 5. Correlation analysis of 2017–2018 data.

Comparison of the correlations is given in Table 4. The higher correlation is found in Assignment, Total visits and Course views. Correlations in first and second year data are close to each other. This is why previous year's data can be used for predicting next year's grades. We can assume that students behaving similarly will receive similar grades.

Table 4. Correlation of attributes with the grade.

Attributes	Correlation coefficients	
	2016–2017 data	2017–2018 data
Total visits	0.55	0.51
Course view	0.43	0.54
Forum view	0.20	0.19
Forum created	0.04	−0.07
Quizzes	0.38	0.68
Assignment	0.69	0.72
File submission	0.69	0.72
Grade view	0.23	0.21

3.2   Prediction

As we found positive correlations between all Moodle activities except the forum creations in second data set; and the course grade as a success measure, we can use this data for modeling. The first file with 2016–2017 data is selected as training set to create the model.

At the first step the model is created by using J48 decision tree algorithm when grades are scaled from 5 to 10. The model is trained with 10 fold cross-validation. Table 5 shows the evaluation of the model with 6 classes for the grade. Overall accuracy of the model is 42.7% while TP rate for the failed students is 68.6%. The model with 6 classes works well for two classes: 5 and 10 only, and could not predict the grade 9 at all.

Table 5. Evaluation of training data set - J48 decision tree with 5 classes.

Class	TP rate	FP rate	Precision	Recall	F-Measure
5	0.686	0.179	0.585	0.686	0.632
6	0.290	0.217	0.295	0.290	0.293
7	0.529	0.230	0.360	0.529	0.429
8	0.000	0.040	0.000	0.000	0.000
9	0.000	0.000	?	0.000	?
10	0.750	0.067	0.545	0.750	0.632
Weighted Avg.	0.427	0.156	?	0.427	?

When the grades are labeled into three classes, such as FAIL for grade 5, GOOD for grades 6–7–8 and EXCELLENT for the grades 9–10; overall accuracy of the model increases to 71.5%. Table 6 shows the evaluation of the model for 3 classes, but TP rate for the FAIL class decreases to 67.1%.

Table 6. Evaluation of training data set - J48 decision tree with 3 classes.

Class	TP rate	FP rate	Precision	Recall	F-Measure
FAIL	0.671	0.105	0.701	0.671	0.686
GOOD	0.824	0.428	0.718	0.824	0.767
EXCELLENT	0.405	0.028	0.739	0.405	0.706
Weighted Avg.	0.715	0.277	0.717	0.715	0.706

When the grades are labeled into two classes, such as FAIL and PASS, overall accuracy of the model increases to 86.5%. TP rate for the FAIL class decreases to 65.7%. Table 7 shows the evaluation of the model for 2 classes.

Table 7. Evaluation of training data set - J48 decision tree with 2 classes.

Class	TP rate	FP rate	Precision	Recall	F-Measure
FAIL	0.657	0.058	0.807	0.657	0.724
PASS	0.942	0.343	0.882	0.942	0.911
Weighted Avg.	0.865	0.266	0.862	0.865	0.861

As it can be seen from Tables 5, 6 and 7; having less classes increase the overall accuracy of the model and precision of the failed grades as well, but the TP value of FAIL class decreases slightly.

We can conclude here that the activities in the learning system are not giving the complete picture of the effort and the grade for the students who passed. But on the other hand it is good measure for the prediction of failed students. For the prediction of real grades, other parameters from the course and the students should be considered, not only online activities. To provide higher overall accuracy for the model, we decided to use 2 classes and we trained the J48 decision tree model with two classes, and got the predictions for the second-year data. Confusion matrix of the test data predictions is given in Table 8. Overall accuracy of the predictions is calculated 89.32% with Kappa index 0.701 for the test data when we compared with the real grade results.

Table 8. Confusion matrix for the predicted data vs. real data with Decision tree algorithm.

Real/Predicted	PASS	FAIL	Classification overall	Precision
PASS	147	10	157	93.63%
FAIL	12	37	49	75.51%
Truth overall	159	47	206	
Recall	92.45%	78.72%		

We applied the same procedure by using Bayesian Network and Support Vector Machine classification algorithms. Tables 9 and 10 show the confusion matrices for the prediction on the test set.

Table 9. Confusion matrix for the predicted data vs. real data with Bayesian Network algorithm.

Real/Predicted	PASS	FAIL	Classification overall	Precision
PASS	146	10	156	93.59%
FAIL	13	37	50	74.00%
Truth overall	159	47	206	
Recall	91.82%	78.72%		

Table 10. Confusion matrix for the predicted data vs. real data with SVM algorithm.

Real/Predicted	PASS	FAIL	Classification overall	Precision
PASS	155	33	188	93.59%
FAIL	4	14	18	77.78%
Truth overall	159	47	206	
Recall	97.48%	29.79%		

In the end Table 11 gives the comparison of the accuracies of three different classification algorithms on the training set and on the test set. The best accuracy is achieved with J48 Decision Tree algorithm (89.32%) on the test set.

Table 11. Comparison of classification methods.

Classification method	Evaluation of the model on the training set		Evaluation of the prediction on the test set	
	Accuracy	Kappa Index	Accuracy	Kappa Index
J48 Decision Tree	86.50%	0.693	89.32%	0.701
Bayesian Network	85.00%	0.624	88.83%	0.69
Support Vector Machine (SVM)	86.15%	0.64	82.04%	0.348

4   Conclusion

This study explains all steps of Moodle log data analysis from raw data to the prediction of new data by using different classification algorithm. Data is taken from a bachelor degree course for two different years. The results show that activities in learning management system have positive correlations with the student success in terms of grade.

The main purpose of the study is to predict students' success and figure out the students who are about to drop out. Differently from many existing studies, we used one year's data to predict the next year's success; so that the drop outs of the students could be recognized in advance.

When the DT model is used with all grades from 5 to 10, it could not predict all the grades accurately. But the most sensitive group of the students (failing ones) with grade 5 were predicted with a high accuracy. Predicting the students who are about to drop out would help instructors to take precautions. Accuracies with passing grades are quite small it means system activity level is not significant on the exam score but it is significant on passing or failing the course. Because of that for predicting all grades accurately learning styles, cognitive styles and study habits of the passing students could be considered in the future studies.

We applied DT model also with three and two labels, and we saw that two labels give the best accuracy overall. By using these two labels we also created BN and SVM models for classification. Accuracy results on the test sets show that DT had the best accuracy on the prediction,

Additionally this study gave us opportunity to try two different tools: RStudio and WEKA which are very popular in data science and to see advantages and disadvantages. We preferred RStudio in pre-processing and visualization as it gives great support for executing SQL queries for data transformation. For the application of decision tree algorithm and prediction we used WEKA as its interface is very simple to set training and test data sets.

In the future, to obtain a model which can predict the grades in a complete scale; different modeling methods can be used. In this study the models were created by using 9 data attributes, in the future more data attributes can be included in the analysis. Also analyzing online activities in a weekly manner and measuring the impact of activities in the first few weeks of the semester could be done as a future work, so the early prediction of drop outs might be useful for further prompt actions.

References

1. Ademi, N., Loshkovska, S.: Exploratory analysis of student activities and success based on Moodle log data. In: 16th International Conference on Informatics and Information Technologies (2019)
2. Akçapınar, G.: Profiling students approaches to learning through Moodle logs. In: Multidisciplinary Academic Conference on Education, Teaching and Learning (MAC-ETL 2015) (2015)
3. Cocea, M., Weibelzahl, S.: Log file analysis for disengagement detection in e-learning environments. User Model. User Adap. Inter. **19**(4), 341–385 (2009)
4. Dumais, S., Jeffries, R., Russell, D.M., Tang, D., Teevan, J.: Understanding user behavior through log data and analysis. In: Olson, J.S., Kellogg, W.A. (eds.) Ways of Knowing in HCI, pp. 349–372. Springer, New York (2014). https://doi.org/10.1007/978-1-4939-0378-8_14
5. Figueira, Á.: Mining Moodle logs for grade prediction: a methodology walk-through. In: Proceedings of the 5th International Conference on Technological Ecosystems for Enhancing Multiculturality, p. 44. ACM (2017)
6. Grothendieck, G.: CRAN - Package sqldf. https://cran.r-project.org/web/packages/sqldf/index.html
7. Kadoić, N., Oreški, D.: Analysis of student behavior and success based on logs in Moodle. In: 2018 41st International Convention on Information and Communication Technology, Electronics and Microelectronics (MIPRO), pp. 0654–0659. IEEE (2018)
8. Käser, T., Hallinen, N.R., Schwartz, D.L.: Modeling exploration strategies to predict student performance within a learning environment and beyond. In: Proceedings of the Seventh International Learning Analytics & Knowledge Conference, pp. 31–40. ACM (2017)
9. Klašnja-Milićević, A., Ivanovic, M.: Learning analytics–new flavor and benefits for educational environments. Inform. Educ. **17**(2), 285–300 (2018)
10. Konstantinidis, A., Grafton, C.: Using Excel macros to analyse Moodle logs. In: 2nd Moodle Research Conference, pp. 33–39 (2013)

11. Poon, L.K.M., Kong, S.-C., Yau, T.S.H., Wong, M., Ling, M.H.: Learning analytics for monitoring students participation online: visualizing navigational patterns on learning management system. In: Cheung, S.K.S., Kwok, L.-F., Ma, W.W.K., Lee, L.-K., Yang, H. (eds.) ICBL 2017. LNCS, vol. 10309, pp. 166–176. Springer, Cham (2017). https://doi.org/ 10.1007/978-3-319-59360-9_15

12. Rachel, V., Sudhamathy, G., Parthasarathy, M.: Analytics on moodle data using R package for enhanced learning management. Int. J. Appl. Eng. Res. **13**(22), 15580–15610 (2018)

13. Estacio, R.R., Raga Jr., R.C.: Analyzing students online learning behavior in blended courses using Moodle. Asian Assoc. Open Univ. J. **12**(1), 52–68 (2017)

14. Romero, C., Ventura, S., García, E.: Data mining in course management systems: Moodle case study and tutorial. Comput. Educ. **51**(1), 368–384 (2008)

15. Romero, C., Ventura, S.: Educational data mining: a review of the state of the art. IEEE Trans. Syst. Man Cybern. Part C (Appl. Rev.) **40**(6), 601–618 (2010)

Multidimensional Sensor Data Prediction

Stefan Popov[1,2] and Biljana Risteska Stojkoska[1(✉)]

[1] Faculty of Computer Science and Engineering, Ss. Cyril and Methodius University
in Skopje, Rudjer Boshkovikj 16, P.O. Box 393, Skopje, North Macedonia
popovstefan@live.com, biljana.stojkoska@finki.ukim.mk
[2] Department of Knowledge Technologies, Jozhef Stefan Institute,
Jamova cesta 39, Ljubljana, Slovenia

Abstract. The ubiquity of modern sensor nodes capable of networking has incentived the development of new and intelligent embedded devices. These devices can be deployed anywhere and by connecting with each other they form a wireless sensor network (WSN). WSN add a new dimension to the world of information which enables the creation of new and enriched services widely applied in different industrial and commercial application areas. The subject of this paper is comparison of different techniques for reduction of overall energy consumption in WSN. Many techniques try to reduce the amount of data sent by the sensor nodes by predicting the measured values both at the source node and at the sink (base station). By doing that, transmission would only be required if the predicted value differs from the measured value by a predefined margin. The algorithms presented here treat the sensor data as part of a time series. They provide great reduction in power consumption and do not require any a-priori knowledge.

Keywords: Wireless sensor networks · Dual prediction scheme ·
Energy saving

1 Introduction

Wireless sensor networks (WSN) generate space and time progressions of the quantities they measure. If their nodes periodically transmit data, then a significant traffic overhead will occur and the network's lifespan will be rather short. To overcome this problem, several techniques for data reduction have been proposed. These techniques aim to reduce the overall traffic load of the network and save substantial amounts of electrical energy.

Data reduction techniques can be divided into three groups: data compression, data prediction and techniques for in-network data processing. This paper deals with the problem of in-network data prediction. The strategy presented here aims to reduce traffic load and save on energy costs while still maintaining low error margins and high rates of accurate predictions. The presented strategy is called dual prediction scheme, because it uses the same model to predict the

© Springer Nature Switzerland AG 2019
S. Gievska and G. Madjarov (Eds.): ICT Innovations 2019, CCIS 1110, pp. 41–51, 2019.
https://doi.org/10.1007/978-3-030-33110-8_4

sensor data both at the source sensor node and at the base station simultaneously. Three different algorithms for data prediction are proposed, simulated and
evaluated against four independent data sets. Results show that this strategy can
save up to 90% of energy. The rest of this paper is organized as follows. The next
section continues by going further in depth with the problem of data processing
and describes the different techniques developed so far. The third section covers
the algorithms proposed. The fourth section gives an overview of the simulation
process and it's results. In the last section these results are analysed and a final
summary is given at the end.

2   Data Prediction Methods

Keeping in mind the limitations of sensor nodes in WSN, the periodical transmission of measured data to the base stations is not a straightforward task. Because
of the limited memory space, the nodes can only save a certain amount of data
and because of their limited computing abilities they cannot compute complex
or long mathematical operations. The biggest issue however, is the energy consumption. In most part, energy is spent by communicating with other nodes.
To increase the nodes' lifespan the communication should be reduced. Many
techniques have addressed this issue, Fig. 1 shows a brief overview.

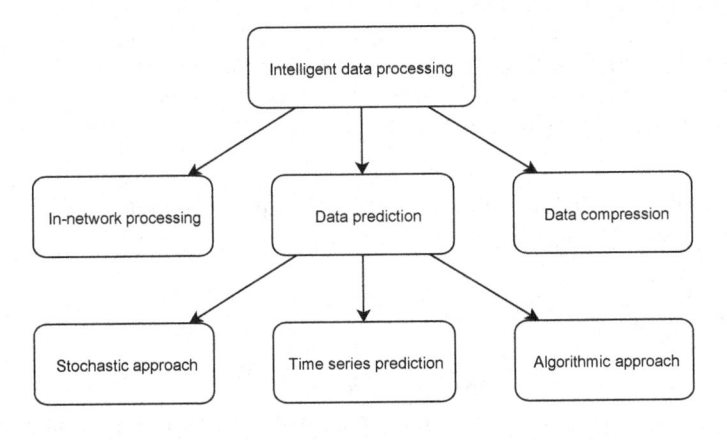

Fig. 1. Overview of data processing techniques

As seen on the figure, the data processing techniques are grouped in three
groups: data compression, data prediction and techniques for in-network processing.

In-network data processing is useful only in situations where the original
measured data values are not needed and an aggregated value can suffice. In-
network processing is executed when the original data arrives at the base station.
The base station can then aggregate (compute minimum, maximum, average,
sum or some other function) the many data values the sensor nodes transmit to it.

By doing that, the base station then communicates only with the aggregated value, thus saving as many transmissions as many original values have been sent.

Data compression techniques use some kind of scheme to encode the sensor data in such a way that the encoding will decrease the overall memory space of the data while still leaving it in such a state that they can be decoded in their original values (or as close to their original value as they could possibly be) in the future. These techniques can be applied when the data is not needed in real time and the encoding and decoding processes are easy to implement. An example of WSN solution utilizing a data compression technique can be found in [10].

Data prediction techniques are most often implemented through the dual prediction scheme. In that scheme the same model (or algorithm, both terms are used interchangeably) is simultaneously being used to predict measured data both at the source node and at the base station. It should be noted that the base station predicts the data for all the sensor nodes it oversees. Both devices should implement the same model so that they execute the same predictions. If the model prediction at the base station is satisfactory i.e. is within a predefined threshold limit, then the predicted value will be accepted and treated as if it was the real, measured value. If however, that is not the case i.e. the prediction differs from the measured value more than the defined threshold limit, then an error has occurred and the real, measured value is transmitted from the source node to the base station. Because both the node and the station execute the same model, both of them will know when an error occurs and a transmission should be expected. Using this scheme, whenever a correct prediction has been made, a node will not transmit a data value and the traffic load shall thus be reduced. Using the same model as the source node, the base station will be able to reconstruct the measured data with a certain precision. Periodically, the models should be validated and updated so that the correctness of a model does not diverge. Once again, keeping in mind the nodes' limitations, a model for data prediction should follow these two guidelines:

1. The model should execute fast, large time complexity is not suitable for the small computing power the nodes posses.
2. The model should also keep track of the memory and should not save many measured values upon which the predictions will be made.

The model complexity will not affect the energy savings as much as it's ability to correctly predict the data value. In [1], chapter 4.2, the authors have conducted experiments for determining the energy consumption of mica2 and mica2dot sensor nodes in different working modes. The results are as follows. Mica2 nodes spent slightly more energy than mica2dot nodes, in all modes. Both nodes show that slightly more energy is needed to send some data (one byte in this particular case) than it is to receive the same amount. Energy consumption during normal CPU mode is two times less than during wireless communication. This is why much research is put into decreasing WSN transmissions. When a node does not send or measure data it should be put in hibernation (sleeping) mode. In

this work mode, the node keeps only one signal line awake on which it waits for a notification about the occurrence of a certain event. Nodes should spend as much time as they possibly can in hibernate mode. If for example, a node only measures and transmits data every 5 min, it should spend the rest 04:59 min hibernating. The amount of energy needed to keep the node in hibernation is 1,000 times lower than the amount needed for it's normal working mode.

Data prediction techniques can be further categorized in techniques with stochastic or algorithmic approach and techniques which treat the data as a time series. The stochastic approach treats the data as a random process and creates a probabilistic model of the data which then is used to predict the future data values. This approach is general and heavyweight for implementation thus it is very rarely used. Algorithmic approach is used whenever a certain heuristic can be used for data prediction and is often application specific. The simplest, most general and easiest to implement, while still maintaining a certain level of accuracy, are the techniques which treat data as time series. The time series models offer great transmission reductions, even in cases where no a-priori knowledge for the data is available.

3   Algorithms for Data Prediction

Times series data prediction in WSN is done in real time and is a great way to reduce network traffic load. In [9] a comparison analysis of three different models is presented. Here, a similar analysis will be introduced, between the following models: simple moving average, single- and double exponential smoothing. The work of these models will be evaluated for different parameter values and on different data sets. These and many other models for time series prediction are described in [6].

3.1   Simple Moving Average

The simple moving average (SMA) model is the easiest to implement as it predicts it's values as just an arithmetic mean of the previous **M** values:

$$x_{n+1} = \frac{1}{M} \times \sum_{i=n-M}^{n} x_i \tag{1}$$

This algorithm is always the first choice for predicting time series data or digital signal processing [8]. It's simplicity has a very good property, i.e. the SMA model offers most reduction of white noise in the data. White noise is a random process i.e. an array of values that are completely random and have no correlation between themselves, and thus they cannot (theoretically) be accurately predicted. SMA is an optimal solution for minimising white noise influence on the prediction process.

3.2   Exponential Smoothing

Predicted time series data consist of two components: prediction function and random error (white noise):

$$x_{t+1} = f(x_n) + \epsilon_n, n = 1, 2, 3, ...t \tag{2}$$

Because ϵ is completely random value and introduces an error which cannot be predicted, the function must offer a solution to minimise it's effect on the prediction. That is done by using a smoothing method. Two exponential smoothing methods will be presented here. An exponential smoothing method breaks the time series data value into multiple components (if they exist): trend, period, cycle, error. The error component is represented with an ϵ in the equation above. Trend is a long term progression in the series, for example the rise of average temperature each year in the last 20 years due to global warming issues. The periodic component consists of a repeating pattern in the values for a particular period or a season. An example would be the monthly temperature each year. The cyclic component represents data patterns which are not periodic or in any way expected but are predictable, for example the temperature will rise near an exploding volcano; the volcano eruption is an irregular occurrence, but the increase in temperature after eruption is expected and can be predicted. Depending on the presence and strength of each of these components, different methods can be used to model them.

Exponential smoothing methods are useful when the time series exhibit presence of periodic and/or trend component. The single exponential smoothing (SES) method assumes that there is not present a periodic component or large variance in recent time series values. The SES method is essentially weighted moving average parameterized with a weighting parameter α which can take on values in the [0, 1] interval. This parameter gives more meaning to the more recently measured values. The most recent measured value is multiplied by a factor of α, and for each step further back in the past a multiplier of $(1 - \alpha)$ is put into effect, so for example, the fourth last measurement will be multiplied by a factor of $\alpha(1 - \alpha)^3$, the fifth last with $\alpha(1 - \alpha)^4$ and so on. This way, an equation for predicting the next measurement in moment t - F_t can be deduced:

$$F_t = \alpha x_{t-1} + \alpha(1-\alpha)x_{t-2} + \alpha(1-\alpha)^2 x_{t-3} + \alpha(1-\alpha)^3 x_{t-4} + ... + \alpha(1-\alpha)^{n-1} x_{t-n} \tag{3}$$

By simplifying this geometric progression, we obtain the following:

$$\begin{aligned} F_t &= \alpha x_{t-1} + (1 - \alpha)[\alpha x_{t-2} + \alpha(1 - \alpha)x_{t-3} + \alpha(1 - \alpha)^2 x_{t-4}] \\ &= \alpha x_{t-1} + (1 - \alpha)F_{t-1} \end{aligned} \tag{4}$$

This next equation shows the final, simplified formulas for data prediction with SES where **F** are the predicted values and **x** are the original series:

$$F_0 = x_0$$
$$F_1 = \alpha x_0 + (1 - \alpha)F_0 = \alpha F_0 + (1 - \alpha)F_0 = F_0$$
$$F_2 = \alpha x_1 + (1 - \alpha)F_1 \tag{5}$$

...

$$F_{t+1} = \alpha x_t + (1 - \alpha)F_t$$

That way, SES model should only memorize two values: α parameter and previously predicted value F_{t-1}. The key point in this model is the value of α. If α is taken to be zero, than the model will "smooth out" all values, and for each **t** the following will be true:

$$F_{t+1} = F_t \tag{6}$$

If however α is to be given a value of one, than for each **t**:

$$F_{t+1} = x_t \tag{7}$$

For any other value between zero and one, α will have different effect on the prediction. Figure 2 illustrates a graphic on which the influence of different α values over the previously predicted data values is shown. It can be noted that the greater α is the greater influence is given to the more recent predicted values and that influence decreases steeper as we go further back in the past.

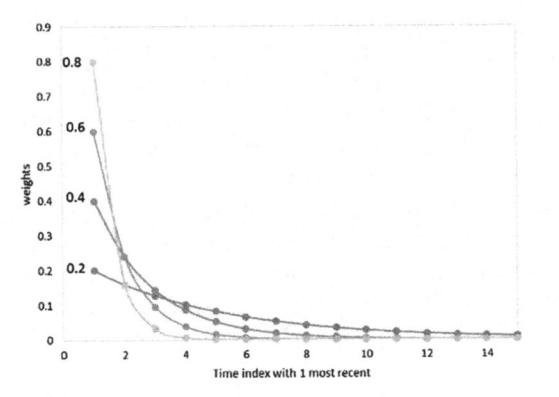

Fig. 2. Influence of α over the weight of previously measured data values. Illustration from [6], page 87

SES does not predict so well when there is periodicity in the data. In order to incorporate the periodical component of the series into the prediction process, the smoothing needs to be done twice, recursively, and this method is called double exponential smoothing (DES). DES takes two parameters α and β (again with values in range [0, 1]) which control the influence of the period component. Predicting with DES is more complex than it is with SES. The predicted value is sum of two terms: periodic component and smoothing factor. DES needs to keep

track of four values: α, β, previous period value T_{t-1} and previous smoothing factor value F_{t-1}. Mathematically, it is represented as follows:

$$\hat{X}_0 = x_0$$
$$F_1 = x_0, T_1 = x_1 - x_0, \hat{X} = F_1 + T_1$$
$$...$$
$$F_t = \alpha x_t + (1 - \alpha)(F_{t-1} + T_{t-1})$$
$$T_t = \beta(F_t - F_{t-1}) + (1 - \beta)T_{t-1}$$
$$\hat{X} = F_t + T_t, t = 2, 3, 4...$$

$$(8)$$

4 Simulation Runs

The algorithms described in the previous section were implemented in Python. The code is available for download at [7]. The basic method around which the whole project is organized is the method **simulate(series, predictedSeries, thresholds)** which simulates the work of a WSN. The method takes three arguments: original (multidimensional) sensor data, the sensor data that has been predicted by applying one of the algorithms over the original series and a matrix of threshold limits. Then, the corresponding elements of the matrices of original and predicted data are subtracted and checked if their absolute difference is within the appropriate threshold limit. As a result the method will return two arrays containing elements of mean square error and percentage of sufficiently correct predictions for each threshold combination. These arrays are further processed to get the final results and evaluate the performance of the algorithms as shown in the figures in the next section. Figure 3 shows a perfect inverse relationship between the values of the two result arrays. Python libraries numpy and matplotlib were used for figure drawing and data processing. The results shown on the figures are used for algorithm evaluation. To obtain those figures, the algorithms were simulated on four real data sets. Those data sets were preprocessed and cleaned. For some measurements a missing value has been noted and ignored because in reality wireless communication is not 100% reliable and package loss can occur.

The first data set (DS1) contains sensor data for air temperature and humidity from 9 rooms in a household generated by a ZigBee sensor network [2]. The set contains 19,735 measurements from each node measured in 3 min interval in a period of 4 and a half months. They were originally used to model and predict energy consumption in the household.

The second data set (DS2) contains meteorological data (temperature, humidity, pressure, PM2.5 concentration levels, speed and wind direction) for five Chinese cities: Beijing, Shanghai, Guangzhou, Chengdu and Shenyang. Data were taken at one hour intervals in the period between 1 January 2010 and 31 December 2015 [4]. This data set has been published as a part of a research project regarding air quality in bigger Chinese cities. Because many of the

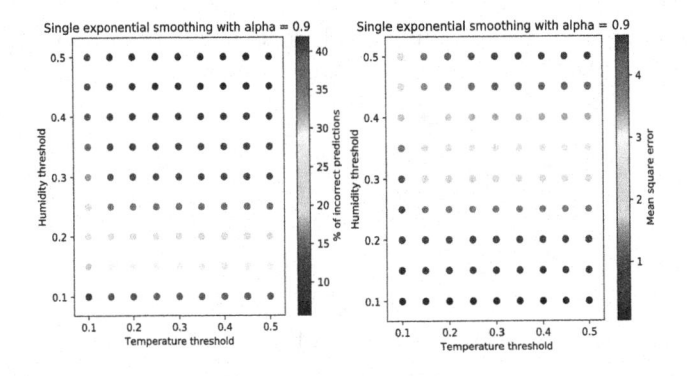

Fig. 3. Simulation run results of applying SES with $\alpha = 0.9$ over DS1. Highly negative correlation between accuracy (left) and error values (right) exists.

measurements are missing for Shanghai, it has been left out of the simulation. The data for temperature, humidity, pressure and dew point will be used only.

Third data set (DS3) was obtained from "Intel Berkeley Research Lab" [3] where 54 mica2dot sensor nodes measuring temperature, humidity and light intensity were deployed. Because many of the measurements are missing for some sensor nodes, in the simulation only the data available from sensor nodes 16, 17, 18, 19 and 20 will be used. Each of them contains circa 35,000 measurements.

The fourth data set (DS4) was downloaded from the national oceanic and atmospheric administration center [5] located near "NASA Johnson Space Center" in Houston, Texas, USA. The center keeps records of atmospheric parameters in the area near by: air temperature, dew point, water currents, salt levels, water surface pressure, water levels and other information regarding the wind and the tides. The simulation executed here will only use the data available for sea level pressure and air and water temperature.

4.1   Results

The following figures show aggregated simulation run results from applying a particular algorithm over all four data sets. For example, DS1 consists of 9 sensor nodes which would generate 9 different accuracy and error values, and in order to visualize them more compactly their values were averaged. This enables us to make more generalized observation of the algorithms' performances. Figure 4 illustrates simple moving average and Fig. 5 the simulations run for the single exponential smoothing method.

Both figures have parameter values on x axis i.e. order value for SMA and decaying factor for SES. Figure 4 shows that for each data set except DS4 a trend in decreasing evaluation metric values with an increasing order can be seen. This particularly applies for DS1 and DS3.

Regarding SES, DS1 and DS4 show similar results in terms of prediction accuracy, but diverge from one another when it comes to error values. DS1 and

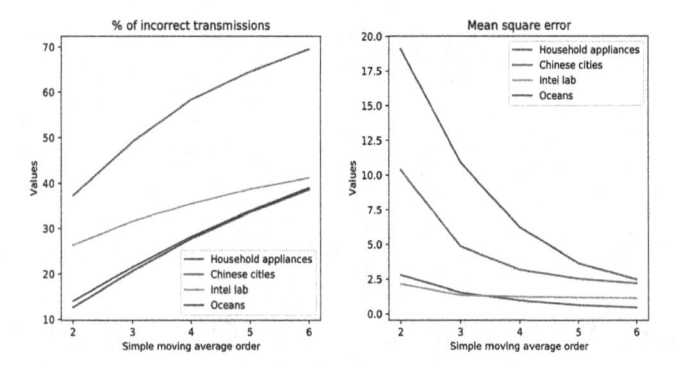

Fig. 4. Simulation run results from applying simple moving average over all four data sets, on the left are the accuracy values and on the right are the error values

DS3 have similar error margins, but different accuracy. DS2 is an outlier from the patterns exhibited from the other three sets (Fig. 6).

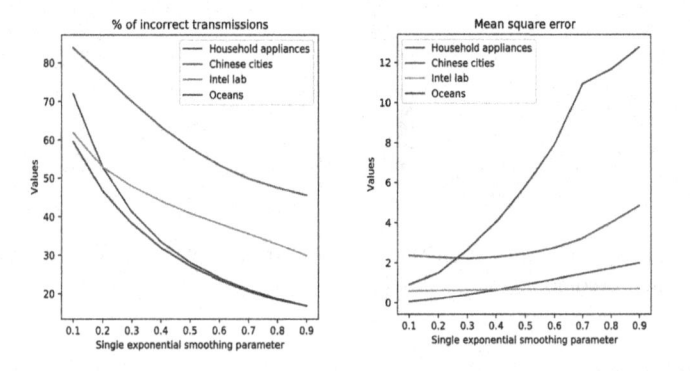

Fig. 5. Simulation run results from applying SES over all four data sets, on the left are the accuracy values and on the right are the error values

DES shows similar patterns as SES. The x axis shows α and β values which are always taken to be the same (both for simpler view and in compliance with previous performance results). In terms of accuracy DES best performs at DS1 and DS4, at DS1 having the least error margin. Other data sets show increasing errors as parameter value also increases, except for 0.9, when it decreases. This suggests that for DES either very small (<0.4) or very large (>0.8) parameter values should be used. Figure 7 shows the effect that α and β have in predicting with DES. The figure shows DES being applied over DS4 with different α and β values on x and y axis respectively. Low β values result in low accuracy and low errors, while high values result in high accuracy and high errors. It can be concluded that the α values do not affect the performance whatsoever. Figure 7 also shows an inverse relationship between accuracy and error values.

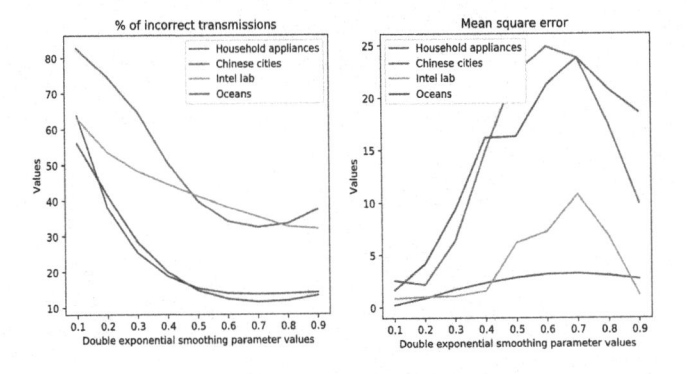

Fig. 6. Simulation run results from applying DES over all four data sets, on the left are the accuracy values and on the right are the error values

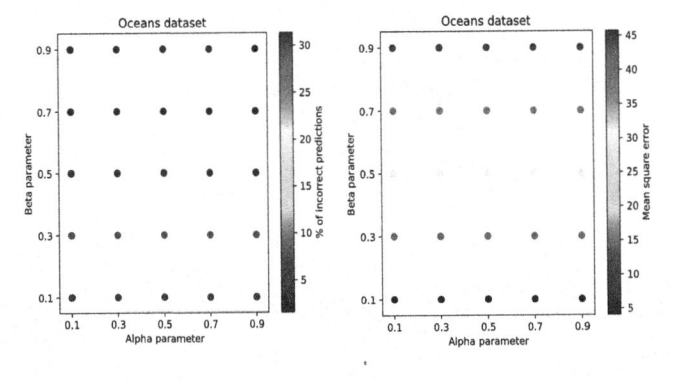

Fig. 7. Simulation run results of DES being applied over DS4. α parameter has almost no effect on the outcome of the prediction

5   Conclusion

This paper dealt with the problem of saving energy in wireless sensor networks by employing dual-prediction scheme for data prediction. A framework for simulation and evaluation of different data prediction algorithm was also developed. Additionally, a comparison analysis for simple moving average, single and double exponential smoothing methods was given. The conclusions are as follows.

As the name suggests, the simple moving average is the simplest one to implement and gives rather predictable results. It performs best when it is used with a small order value (<4). Single exponential smoothing method keeps track of only one past predicted value and it's α parameter. The bigger α is the better the results are and a value of ($0.7 < \alpha < 1$) is highly suggested. Double exponential smoothing method is somewhat complex to implement and requires much more calculations than the previous two models. It however still keeps track of only two previous values and two parameters. It was demonstrated that only the β parameter has an effect on the simulation results, but in isolated

simulations it was shown that α can in small doses still effect the outcome and a value very close to β is always recommended. Also, DES shows great results when very small (<0.4) or very large (>0.8) parameter values are used.

From these analysis it is somewhat hard to say that the performance of the algorithm depends on the nature of the data, even though there are very clear patterns that support that argument. Figure 4 most clearly shows that DS1, DS3 (indoor data) and DS2, DS4 (outdoor data) are clustered in terms of both accuracy and error values. These clusters are also noticeable in the smoothing methods' simulation runs, but they diverge from one another in terms of accuracy. For a future WSN application in an outdoor data setting, we would recommend using simple moving average with a low order value (<4) because it most easily deals with the high fluctuations and variance this type of data most often exhibits, and for applications regarding indoor data settings, either SES or DES with large parameter values (>0.8) because they require less memory and can easily model the slowly changing, low variance phenomena.

References

1. Anastasi, G., Falchi, A., Passarella, A., Conti, M., Gregori, E.: Performance measurements of motes sensor networks. In: Proceedings of the 7th ACM International Symposium on Modeling, Analysis and Simulation of Wireless and Mobile Systems, pp. 174–181 (2004)
2. Candanedo, L.M., Feldheim, V., Deramaix, D.: Data driven prediction models of energy use of appliances in a low-energy house. Energy Build. **140**, 81–97 (2017)
3. Intel Lab data. http://db.lcs.mit.edu/labdata/labdata.html. Accessed 1 Sept 2019
4. Liang, X., Li, S., Zhang, S., Huang, H., Chen, S.: PM2.5 data reliability, consistency, and air quality assessment in five Chinese cities. J. Geophys. Res.: Atmos. **121**, 6742–6761 (2016)
5. National Oceanic Atmospheric Administration's National Data Buoy Center, Historical data. https://www.ndbc.noaa.gov/download_data.php?filename=41024h2018.txt.gz&dir=data/historical/stdmet/. Accessed 1 Sept 2019
6. Pal, A., Prakash, P.: Practical Time Series Analysis, 2nd edn. Packt Publishing, Birmingham (2017)
7. Python Jupyter notebook with the code for this project. https://github.com/popovstefan/undergraduatethesis. Accessed 1 Sept 2019
8. Smith, S.W.: The Scientist and Engineer's Guide to Digital Signal Processing, 2nd edn. California Technical Publishing, Poway (1999)
9. Stojkoska, B.R., Mahoski, K.: Comparison of different data prediction methods for wireless sensor networks. In: Proceedings of the 10th Conference for Informatics and Information Technology, pp. 307–311 (2013)
10. Stojkoska, B.R., Nikolovski, Z.: Data compression for energy efficient IoT solutions. In: Proceedings of the 25th Telecommunication Forum, pp. 1–4 (2017)

Five Years Later: How Effective
Is the MAC Randomization in Practice?
The No-at-All Attack

Ivan Vasilevski[✉], Dobre Blazhevski, Veno Pachovski,
and Irena Stojmenovska

University American College Skopje, Skopje, Republic of North Macedonia
vasilevskiivan@gmail.com, {dobre.blazevski,pachovski,
irena.stojmenovska}@uacs.edu.mk

Abstract. The user privacy, in particular, user tracking, has always been a considerable concern, and moreover nowadays, when we are completely surrounded by Wi-Fi enabled devices (smartphones, tablets, wearables, etc.). These devices transmit unique unencrypted signals containing information which includes a device's MAC (Media Access Control) address. Such signals can be monitored with a passive attack by using cheap hardware. Since the MAC address is unique for each device, there is an unquestionable privacy threat to the devices' owners. To this moment, the only countermeasure vendors have the MAC Address Randomization. In this paper, we show that the effectiveness of this solution, five years after it was introduced for the first time, is insufficient to prevent Wi-Fi users from tracking. Moreover, the solution itself is not even widely used.

To validate such conclusions, we have conducted a week-long passive attack using Single Board Computer (Raspberry PI), and we were able to obtain real-world sample data (7.522 total wireless probe requests). Thus, we were being able to: analyze data and count users, notify their presence measure time they spent in an area, determine the working hours and the busiest day, distinguish vendors, etc.

The paper also suggests mitigations, including some that may affect the MAC Randomization implementation itself as well as the user behavior.

Keywords: MAC randomization · Wi-Fi · Passive attack · Privacy · Network · Single Board Computer

1 Introduction

According to [1], Wi-Fi is the most commonly used wireless communications technology and the primary medium for global Internet traffic with 13 billion devices in use in 2018. It is based on the Institute of Electrical and Electronics Engineers (IEEE) wireless communication standard 802.11 and is present in every mobile device, such as mobile phones, laptops, tablets, etc. Since every Wi-Fi enabled device is using an MAC address, and the MAC address is unique for each device, there is a possibility to track the owner of the device [2]. Some companies are using Wi-Fi to track users in the

© Springer Nature Switzerland AG 2019
S. Gievska and G. Madjarov (Eds.): ICT Innovations 2019, CCIS 1110, pp. 52–64, 2019.
https://doi.org/10.1007/978-3-030-33110-8_5

shops in order to obtain shopping behaviors [3–5], whereas examples for tracking people in urban areas exist as well [6]. However, tracking users raises a significant privacy issue. The uniqueness of mobility traces is so high that only four spatiotemporal points are sufficient to identify 95% of individuals in a population of 1.5 million individuals [7]. In [8], MAC randomization was proposed as a countermeasure for Wi-Fi tracking by using a disposable identifier instead of a device's MAC address. Vendors started implementing MAC address randomization in 2014 [9]. However, since then, different attacks appeared capable to defeat the randomization.

In our work, we are investigating how effective MAC randomization is in practice, five years after the first implementation efforts. For that purpose, we conducted a passive attack on the IEEE 802.11 by using Raspberry PI with software for monitoring probe requests. Inspired by the approach in [10], we are analyzing the IEEE 802.11 standard's probe requests in order to see whether there are random MAC addresses used by mobile devices during the discovery of AP (Access Points). For the completeness of the paper, we firstly make a literature review referring to the basic principles of the 802.11 standard device discovery and the MAC randomization. Then we present related work in the field, as well as the attacks on the MAC randomization, is known so far. Afterward, we explain in detail our methodology and findings. The final section contains a conclusion - based on our results and the literature review given, including some suggestions for possible improvements to this concept and the user privacy in general.

2 Background

2.1 802.11 Standard Device Discovery

When using an 802.11 wireless network (Wi-Fi), before joining the network, mobile stations must identify whether that network is compatible. As it is specified by Gast [11, 12], the process of identification of existing networks is called scanning. The case when a mobile station only listens for a beacon frame sent from the Access Points (AP) in order to discover them, is called passive scanning. In order to perform active scanning for a wireless network, Wi-Fi-enabled mobile stations use Probe Request frames to scan an area for existing Wi-Fi networks [11] (Fig. 1).

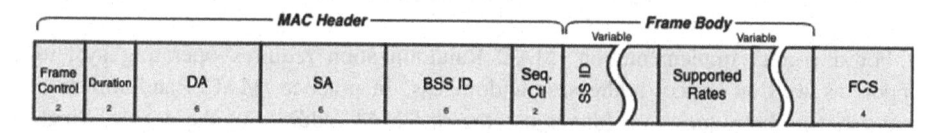

Fig. 1. Probe request frame.

Probe Request frame contains information about network capability, RSSI (Received Signal Strength Indicator) signal, previous SSID (Service Set Identity) [11, 13]. Most importantly, it contains the mobile station's MAC address. After sending the Probe Request, the mobile station starts its Probe Timer countdown. During the

countdown period, it waits for some answers (Probe Response) from AP called. At the end of the timer countdown, the mobile station processes the answer it has received.

However, if no answers received, the mobile station moves to the next Wi-Fi channel and sends another probe request. Hence, Probe Requests are sent without being connected to the Wi-Fi network [14].

On the other hand, the beacon frame, which is one of the management frames in IEEE 802.11 based WLANs, is sent periodically at a time called Target Beacon Transition Time (TBTT) in a time unit (TU) - which is a unit of time equal to 1024 microseconds. By default, the beacon is transmitted every 100TU (102, 400 µs) [11].

2.2   MAC Randomization

In order to communicate, every device possesses a MAC address. The MAC address is considered globally unique. In order to ensure the global uniqueness of MAC addresses, the first three bytes (prefix) of this address is an Organizationally Unique Identifier (OUI) which has to be bought by vendors from the IEEE Registration Authority in order to be used (see Fig. 2). The last three bytes are the Network Interface Controller (NIC) bytes.

As shown in Fig. 2, one particular bit of interest in the MAC address is the Universal/Local (U/L). If this bit is set to 1, then it indicates that the MAC address has been changed by the administrator of the device and is not guaranteed to be unique as from the factory, but is changed to a local MAC address instead [15]. On the other hand, this may also be used to create randomized MAC addresses as a measure of privacy [10].

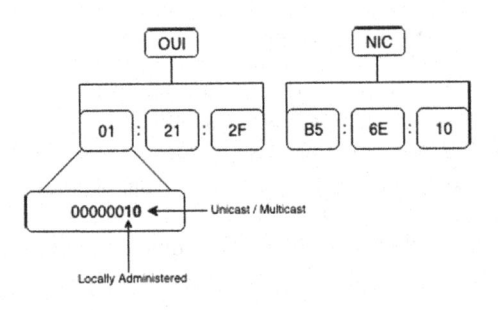

Fig. 2. 48-bit MAC address structure

For a correct implementation, MAC Randomization requires operating system support as well as correct addresses randomness. In practice, MAC Randomization means that probe requests no longer use the real MAC address of the device. Since there is no specification on MAC address randomization, Windows, Linux, Android, and iOS all have implemented their variants of randomization [10, 16–20].

On the Windows Hardware Engineering Community conference (WinHEC), Microsoft announced MAC Randomization on Windows 10 as a feature [16], enabling randomization depending on the hardware and driver support. To ensure randomization, Windows uses random addresses for probe requests and also uses random addresses when connecting to a network [18, 21].

MAC Randomization implementation on Windows 10 is not perfect as there are still ways to bypass it, but when enabled, it significantly improves the end-user privacy [22].

Linux implemented MAC Address randomization during network scan since kernel version 3.18. In Linux, the Wi-Fi driver generates per burst random MAC Addresses. Linux has a default duration of 60 s for a random MAC address located in wpa_supplicant [23]. Linux has 3 MAC address policies: use permanent MAC address, use random MAC address, use random MAC address with maintained OUI. The way that the MAC randomization is implemented in Linux makes it vulnerable as such [10].

In early 2015, Google introduced the MAC Address randomization as a feature that was intended to increase user privacy and avoid real-time tracking [17], but it is not used consistently by all manufacturers [10]. Google's solution was to rotate a sequence of pseudo-random addresses while scanning for networks and revealing the real MAC Address, once the device is connected to the network. The solution features periodical update of the MAC address to a random value, but most Android devices simply do not have this feature enabled, even though the majority of them has versions of the operating system that should allow using it. Troublesomely, Android devices that have this feature enabled tend to reveal the hardware real MAC Address regularly, even when they are not associated with the network. As stated in [10], in most of the cases, this happens when the device is receiving a call, and when the device is turning on its screen. On the other hand, Android devices can be forced to reveal the MAC address even when the user disables Wi-Fi or enables Airplane Mode, in the case when location-service is enabled [10].

With iOS 8, Apple introduced the newly implemented MAC Randomization feature [24]. At first, they offered MAC Randomization only when the device is not connected to any network while the device is in sleep mode [18, 19]. The original address is used when the device connects to a network [18]. As it is shown in [20], later in iOS 9, the MAC Randomization feature was also extended to use randomization in location and auto-join scans, which is used for automatic joining to open Wi-Fi hot-spot. Interestingly Apple devices do not transmit Wi-Fi Protected Setup (WPS) fields to indicate any other sort of model information other than the iOS version [10].

3   Related Work

After the implementation of MAC Randomization, researchers presented several attacks on how randomization can be defeated. It was shown that the MAC address is not the only information that can be used for tracking.

One approach was presented in [25] and is related to iOS devices. Namely, the sequence number field of probe requests was not reset when the MAC address was changed. This was used for further device tracking.

Then, as it is showed in [23], during active scanning, mobile stations tend to follow predictable temporal patterns of probe requests, which are sent in short bursts. Therefore, temporal fingerprints can be created in order to isolate and track the device. This is done by measuring the temporal distribution of probe requests within a burst and between bursts. Timing attack that can defeat MAC randomization and therefore enable device tracking was presented in [26] as well.

Researchers in [22] suggest another way to defeat MAC randomization. Probe requests contain so-called Information Elements (IE) in their payload, which contains information about the capabilities supported by the mobile station. Such information is enough to track some device. One IE they point out is WPS IE as part of the probe request, which can allow reverting back to the original MAC address of some device. Another element they point out is SSID, which is used to specify a network searched by the device and can be used to identify and track some device. There is also an attack that advertises SSIDs in order to get association requests from mobile devices known as Karma attack [27]. This attack uses a fake AP and relies on the SSIDs that the mobile device broadcasts as part of the probe requests and therefore can allow device tracking.

In [22], they even discovered a predictable field located at the physical layer, which allows linking together random MAC addresses of the same device. As a result, although using different MAC addresses, the device can be locked and tracked.

As mentioned in the introduction part, in our research, we are using the approach of Martin et al. [10], in identifying whether there is MAC randomization. They performed broad capture and analysis of Wi-Fi traffic in order to determine which device models and operating systems are using MAC randomization. Martin et al. [10] were using U/L bit as a starting point to identify MAC randomization. We took in consideration their findings regarding known prefixes (such as Google's DA:A1:19 and Motorola's 92:68: C3) and regarding iOS's randomization that is completely random including the 24 OUI bits, where the local bit was always set to 1.

4   Research Methodology

In continuation, we present our research equipment design, data gathering, data collection procedures, results obtained, and data analyses.

The initial objective is identifying if there are devices that are using MAC Address randomization. During a weeklong survey, we captured 802.11 device probe requests traffic (which is unencrypted), by using inexpensive commodity hardware and open-source software.

The experimental setup was established on a semi-public place (premises with a certain number of employees which are providing services to visitors) where we set our equipment. Measures were taken in order not to collect other data than needed. Therefore, for the purpose of our research, i.e., in order to show the effectiveness of the MAC randomization, our setup did not capture actual user traffic, but only Probe Requests transmitted from mobile devices. In our case, there were no MAC addresses that were used for P2P services and Wi-Fi extenders.

4.1   Equipment and Software

We primarily used a single-board computer (SBC) – "Raspberry Pi" with installed Raspbian, a Debian based operating system for the device, and an individual Ralink RT5572 USB wireless card. The software that we used is a simple Python script designed for the purpose of this research. The "tshark" application was also used for the scanning process.

4.2 Equipment Setup

The Ralink RT5572 USB wireless card is connected to the SBD USB port. Our Python script uses the card for the purpose of scanning. "Ralink RT5572" is capable of working in promiscuous mode, and therefore we can obtain passive scanning, i.e., nothing is being transmitted from our wireless card. The script scans the 2.4 GHz band with a specialized tool called "tshark". Then it reads the "tshark" output data and sorts it according to our needs. The collected data is sorted and kept in JavaScript Object Notation (.JSON) file on the server that makes it easily accessible via web browser or triggered via hooks.

The set up was capable of collecting packets on a 2.4 GHz band. There is a total of 14 channels in the 2.4 GHz band. In order to have successful data gathering, a full scan is needed, which will take approximately 3 s or 102,400 μs × 2 × 14 = 2,867,200 μs [28]. In particular, our setup scans in three intervals, where each interval is 3 s long (total 9 s) and then writes collected data in a file.

4.3 Questionnaire

In order to achieve the objective of our research, we were also interested in whether other people have some knowledge of MAC randomization. To get a quick result, we performed simple research using a questionnaire method where we collected 67 responses. In the interest of conciseness, we asked two questions only:

1. Are you familiar with the term "MAC address randomization"?
2. Are you using MAC address randomization on some of your portable devices?

4.4 Results and Analysis

In this section, we present results and evaluate how much information can be revealed. With an efficient process, we ignored duplicate sample polls in order to avoid total user duplication.

A total of 67 people took part in the questionnaire (see Table 1). There is no data obtained about sex and age.

Table 1. Questionnaire results

Question	Answer	
	Yes	No
Are you familiar with the term MAC address randomization?	28	39
Are you using MAC address randomization on some devices?	9	59

We can conclude that 58% of the respondents are NOT familiar with the term "MAC address randomization" and that only 8% replied that they use MAC address randomization on some of their devices. However, these results indicate that the majority of people are not familiar with MAC randomization, whereas very few are actually using it.

As shown in Table 2, 7,522 probe requests were captured during the research period. From those 7,522 requests, we were able to extract 671 MAC addresses.

Table 2. Number of probe requests and MAC addresses during research

Dataset	Raspberry Pi
Mac addresses	671
Probe requests	7,523
Time frame	Seven days

The next step was to determine how many of the extracted MAC addresses are randomized, so we would be able to count how many of the devices were using MAC randomization feature. In order to do that, we checked the U/L bit of every MAC address captured. Parts of captured MAC addresses with their OUI parts are shown in Fig. 3.

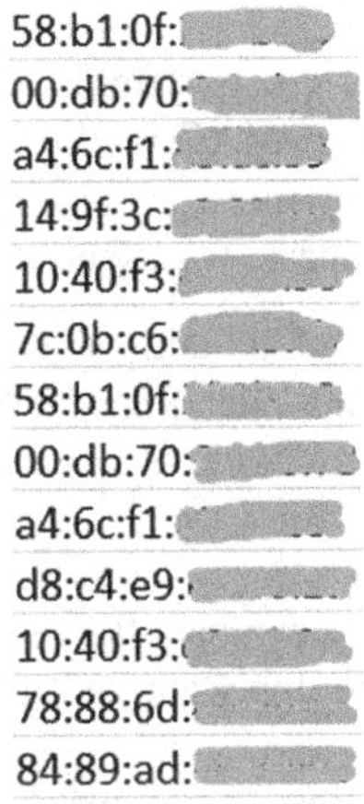

| 58:b1:0f: |
| 00:db:70: |
| a4:6c:f1: |
| 14:9f:3c: |
| 10:40:f3: |
| 7c:0b:c6: |
| 58:b1:0f: |
| 00:db:70: |
| a4:6c:f1: |
| d8:c4:e9: |
| 10:40:f3: |
| 78:88:6d: |
| 84:89:ad: |

Fig. 3. OUI part of captured MAC addresses

After analyzing the U/L bits, we arrived at the finding that captured MAC addresses had no U/L bit enabled. Prefixes such as DA:A1:19 or 92:68:C3 were not detected. This leads towards the conclusion that MAC Randomization feature was not used by the devices. This means that all captured MAC addresses are unique and therefore, 671 unique devices were identified. Knowing this, we were able to proceed with the analysis and to obtain more results.

The semi-public place has working hours from 08:00 to 20:00 on Monday-Friday. Figure 4 shows the number of gathered probe requests between two working days. It can be noticed that the number of probe requests starts increasing as the semi-public place starts to work. Then, the number of probe requests decreases by the end of the working hours, and when the working hours are over a couple of devices are active only. Later we determined that these are some office Wi-Fi devices.

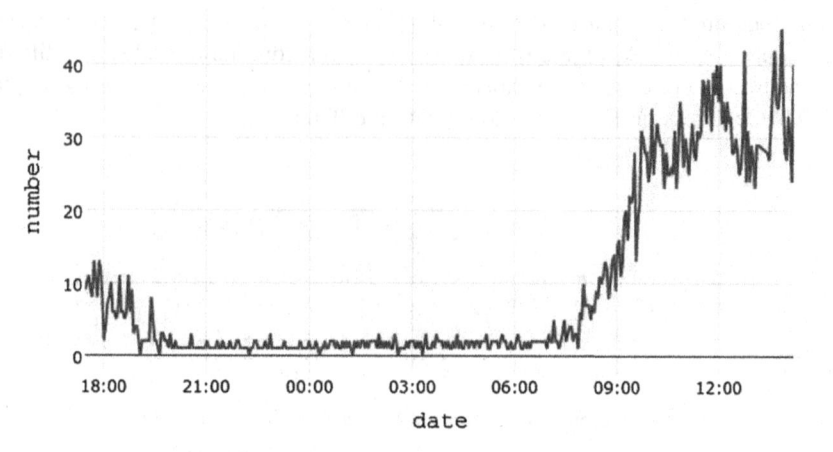

Fig. 4. Number of probe requests related with working hours

Figure 4 shows that by having such measurements, it would be obvious to determine working hours, even if the researcher did not know the working time. Having such results in mind, we analyzed and counted probe requests per day during the week (see Fig. 5).

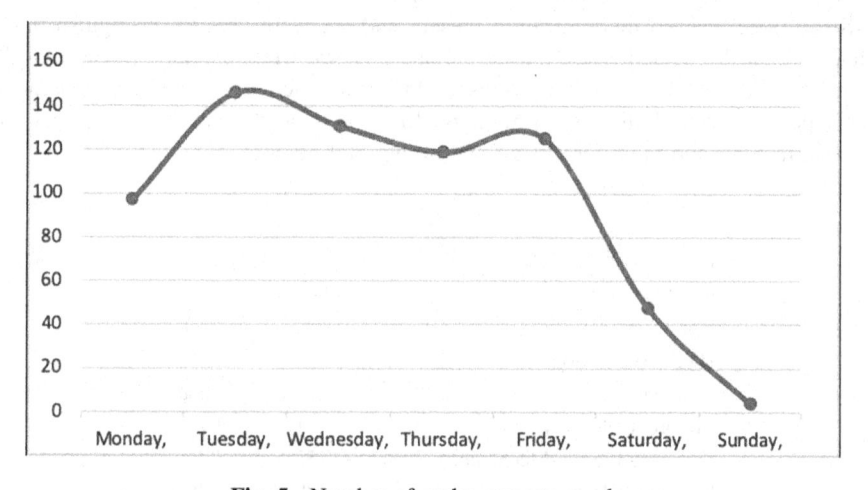

Fig. 5. Number of probe requests per day

As a result, it can be determined that although the regular working time of the place is Monday to Friday, there are some devices active even on Saturday. However, it can be definitely concluded that there is no activity, i.e., no visitors on Sunday. By analyzing Probe Requests per day, we were also able to determine that Tuesday is the most frequent day at the research site with 146 unique MAC Addresses captured.

The measured and analyzed results showed that the majority of MAC addresses repeats during the week. Figure 6 shows one segment from these results, i.e., different MAC addresses presence at the experimental area for about one hour during the day. Each MAC address on Fig. 6 is colored with a different color.

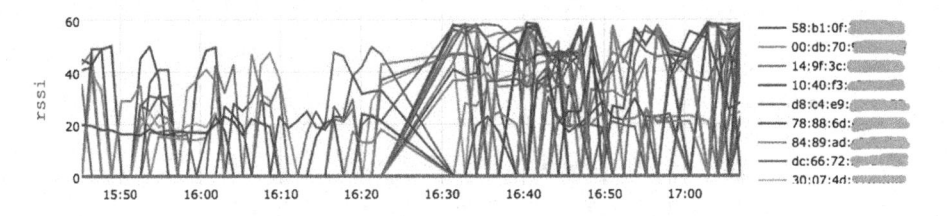

Fig. 6. Presence of different MAC addresses at research site

Figure 6 clearly shows how one MAC address/device travels, i.e., approaches and leaves (movements) from the experimental area during the specified period. The presence is determined through the RSSI (Received Signal Strength Indicator) level obtained for each MAC address/device. Therefore, we were able to determine when some MAC address/device is present in the area and how far it is from the AP. It is essential to mention the comparable data, i.e., during such movement, there was no indication showing MAC randomization at any time.

Since there was no randomization present and we were able to distinguish individual original MAC addresses, during the analysis, we were also able to connect all collected MAC addresses to the corresponding manufacturer. Randomly generated MAC addresses would affect such findings. Table 3 shows the number of devices by the manufacturer and their percentage within the total number.

Table 3. Number of devices by manufacturer

Manufacturer	Devices	%
Apple, Inc.	120	17,9%
HTC Corporation	14	2,1%
LG Electronics	17	2,5%
Motorola Mobility LLC	21	3,1%
Samsung Electro-Mechanics (Thailand)	52	7,7%
Samsung Electronics Co. Ltd	428	63,8%
Xiaomi Comm. Co. Ltd	19	2,8%

5 Discussion and Conclusion

The goal of this paper was to show the effectiveness of MAC Randomization as a countermeasure for tracking users over Wi-Fi, five years after the first implementation efforts. For that purpose, we conducted a passive attack, which allowed us to collect

probe requests transmitted from mobile stations. Inspired by an approach of Martin et al. we analyzed IEEE 802.11 standard's probe requests in order to see whether there are random MAC addresses used by mobile devices during the discovery of AP (Access Points). By using literature review, first, we referred to the most basic principles of the 802.11 standard device discoveries and to the MAC randomization. We also presented related (literature reviewed) work in the field and known attacks on MAC randomization. Different vendors and operating systems implement randomization differently, and therefore, different weaknesses can be exploited. There are known methods/attacks that can defeat MAC randomization, such as Karma attack, by using IE fields in probe requests (WPS and SSID), by using different timing attacks, by using sequence number and by using field located at the physical layer. Moreover, there are cases with Android devices where real MAC address is revealed when enabling screen/receiving a call, or even when Wi-Fi is not enabled.

The results we have obtained in our research were surprisingly unexpected. Having in mind that researchers already discovered that the MAC randomization is implemented by different vendors and that some small percentage of people who participated in the questionnaire were using it, our expectations were to see some percent of devices not using the randomization.

However, after analyzing known prefixes and the U/L bit of MAC addresses we collected, we were able to see that none of the devices used MAC randomization (related work and literature review did not point out complete randomizations where U/L bit is set to 0).

As a limitation to our research, we could point out the small number of captured time frames and extracted MAC addresses in a limited period of time and space. However, having in mind that no randomization was detected at the specified time period, we concluded that there was no need to collect probe requests from different locations at the premises (semi-public place) where APs are also available. Namely, it would be expected that we would receive the same results at different locations at that specified period of time.

Since there was no randomization in use, we were able to determine the working hours of the semi-public place by observing when the number of probe requests increases and decreases. We were also able to precisely determine when people are present at the premises during the week and what the most frequent day in the week is. Such information could give attackers a very good idea about when is right time to perform different attack activities.

By cross-referencing all the MAC addresses with the correspondent timestamp, it is obvious to determine that whenever some device is in the area, there would be probe requests from that device. When the device moves out of range, we will stop receiving probe requests, so it will be safe to assume that the person who carried that device left the area and therefore its movement can be tracked.

We were also able to connect MAC addresses with manufacturers. This can also be very beneficial for attackers. By having such information, attackers would be able to target specific device/users with a specific approach.

Since there is no randomization used, many other data could be extracted by using analysis. In this paper, we presented only some of them.

According to our findings from the literature review, Windows 10 OS offers the best implementation of MAC randomization so far. During the time period of our research, we did not have a chance to detect randomization from a device using Windows 10, although we knew there were some laptops in the premises using Wi-Fi during the research. This means that users of these devices did not turn on MAC randomization.

Probe Requests can be transformed into a fingerprint, i.e., can be used to classify the MAC address uniquely. Having in mind the vulnerabilities presented in this paper and our findings, it is more than evident that MAC randomization only makes user tracking harder, but not impossible. Through our research, we were able to prove that MAC Randomization is not even in use, which leads to the conclusion that there is *not-at-all attack* needed in order to track users and to extract different data that some attacker needs. Having this in mind, we are encouraged to perform the same research on different and busier sites, as part of our future work, in order to gather data on a large scale about MAC randomization usage.

As a summary of all of the above as well as our investigations mentioned so far, we give several suggestions for the future development of the concept.

1. First of all, if implemented, MAC randomization should be enabled by default. However, after enabling Wi-Fi on the device, users shall be notified about some obvious downsides of MAC randomization use and therefore give them an option to disable the randomization if needed.
2. Due to the fact that there are proven cases when Android devices reveal their MAC addresses when enabling screen/receiving a call, or even when Wi-Fi is not enabled, vendors must not allow such a situation. In other words, if some option is not enabled by the user (in this case Wi-Fi), it should not be enabled automatically, without user approval. Otherwise, it presents a severe threat on security.
3. No matter that almost half of the users in our questionnaire stated that they know what MAC randomization is, the results from our research showed that users were not using it at all. (This means that the users themselves are their own greatest 'enemies'). Hence, user education and awareness are necessary, no matter how much MAC randomization is safe or not since it is the only protection available at the moment.

References

1. Wi-Fi Alliance: Wi-Fi Generations (2019). https://www.wi-fi.org/discover-wi-fi. Accessed 13 Feb 2019
2. Cunche, M.: I know your mac address: targeted tracking of individual using Wi-Fi. J. Comput. Virol. Hacking Tech. **10**(4), 219–227 (2014). ISSN: 2263-8733
3. Goodin, D.: No, this isn't a scene from minority report. This trash can is stalking you. Ars Technica (2013). https://arstechnica.com/information-technology/2013/08/no-this-isnt-a-scene-from-minority-report-this-trash-can-is-stalking-you/. Accessed 12 Feb 2019
4. Aerohive. http://docs.aerohive.com/pdfs/AerohiveSolution_Brief-Retail-Analytics.pdf. Accessed 03 Dec 2017

5. Fung, B.: How stores use your phone's WiFi to track your shopping habits (2013). https://www.washingtonpost.com/news/theswitch/wp/2013/10/19/how-stores-use-your-phones-wifi-to-trackyour-shopping-habits/. Accessed 12 Mar 2016

6. Musa, A.B.M., Eriksson J.: Tracking unmodified smartphones using Wi-fi monitors. In: Proceedings of the 10th ACM Conference on Embedded Network Sensor Systems, Sen-Sys 2012, pp. 281–294. ACM, New York (2012)

7. de Montjoye, Y.A., Hidalgo, C., Verleysen, M., Blondel, V.: Unique in the crowd: the privacy bounds of human mobility. Sci. Rep. 3(3), 1376 (2013). https://doi.org/10.1038/srep01376

8. Gruteser, M., Grunwald, D.: Enhancing location privacy in wireless LAN through disposable interface: a quantitative analysis. Mobile Netw. Appl. 10(3), 315–325 (2005). https://doi.org/10.1007/s11036-005-6425-1

9. Cimpanu, C.: Researchers Break MAC Address Randomization and Track 100% of Test Devices (2017). https://www.bleepingcomputer.com/news/security/researchers-break-mac-address-randomization-and-track-100-percent-of-test-devices/. Accessed 23 Dec 2018

10. Martin, J., et al.: A study of MAC address randomization in mobile devices and when it fails. Proc. Priv. Enhancing Technol. 4, 365–383 (2017)

11. Gast, M.S.: 802.11 Wireless Networks: The Definitive Guide, 2nd edn. O'Reilly Media, Inc., Sebastopol (2005)

12. Gast, M.S.: 802.11 Wireless Networks: The Definitive Guide. O'Reilly Media, Inc., Sebastopol (2002)

13. Nadeem, M.: How we tracked and analyzed over 200,000 people's footsteps at MIT. freeCodeCamp (2017). https://medium.freecodecamp.org/@moinnadeem. Accessed 22 Feb 2019

14. Miorandi, D.: I lie, you lie, everybody lies: WiFi tracking in the era of MAC randomization (2017). https://medium.com/@DanieleMiorandi/i-lie-you-lie-everybody-lies-wifi-tracking-in-the-era-of-mac-randomization-2ab147857b24. Accessed 23 Jan 2019

15. Loveless, A.: MAC Address: Universally or Locally Administered Bit and Individual/Group Bit (2011). https://packetsdropped.wordpress.com/2011/01/13/mac-address-universally-or-locally-administered-bit-and-individualgroup-bit/. Accessed 20 Dec 2018

16. Huitema, C.: Experience with MAC address randomization in windows 10. In: IETF 93 Proceedings. IETF, Prague (2015). https://www.ietf.org/proceedings/93/slides/slides-93-intarea-5.pdf. Accessed 05 Jan 2019

17. Goodin, D.: Shielding MAC addresses from stalkers is hard and Android fails miserably at it (2017). https://arstechnica.com/information-technology/2017/03/shielding-mac-addresses-from-stalkers-is-hard-android-is-failing-miserably/. Accessed 19 Jan 2019

18. van Hoef, M.: How MAC Address Randomization Works on Windows 10 (2016). http://www.mathyvanhoef.com/2016/. Accessed 05 Jan 2019

19. Beasley, M.: More details on how iOS 8's MAC address randomization feature works (and when it doesn't) (2014). https://9to5mac.com/2014/09/26/more-details-on-how-ios-8s-mac-address-randomization-feature-works-and-when-it-doesnt/. Accessed 14 Jan 2019

20. IT Dojo: An exploration of iOS' MAC randomization feature and how it works in practice with Apple's iOS 9 (2016). https://www.youtube.com/watch?v=dxj7bqe7wag. Accessed 09 Jan 2019

21. Microsoft Support: How and why to use random hardware addresses (2017). https://support.microsoft.com/en-us/help/4027925/windows-how-and-why-to-use-random-hardware-addresses. Accessed 05 Feb 2019

22. Vanhoef, M., Matte, C., Cunche, M., Cardoso, L.S., Piessens, F.: Why MAC address randomization is not enough: an analysis of Wi-Fi network discovery mechanisms. In: Proceedings of the 11th ACM on Asia Conference on Computer and Communications Security, pp. 413–424. ACM, New York (2016)

23. Matte, C., Cunche, M., Rousseau, F., Vanhoef, M.: Defeating MAC address randomization through timing attacks. In: ACM WiSec 2016. Darmstadt, Germany (2016). https://doi.org/10.1145/2939918.2939930, https://hal.inria.fr/hal-01330476/document. https://arxiv.org/abs/1703.02874 Accessed 14 Jan 2019

24. Apple: About the security content of iOS 8 (2017). https://support.apple.com/en-us/HT201395. Accessed 04 Feb 2019

25. Freudiger, J.: How talkative is your mobile device?: An experimental study of wi-fi probe requests. In: Proceedings of the 8th ACM Conference on Security & Privacy in Wireless and Mobile Networks, p. 8. ACM, New York (2015). https://doi.org/10.1145/2766498.2766517

26. Desmond, L.C.C., Cho, C.Y., Pheng, T.C., Lee, R.S.: Identifying unique devices through wireless fingerprinting. In: Proceedings of the First ACM Conference on Wireless Network Security, pp. 46–55. ACM, New York (2008). https://doi.org/10.1145/1352533.1352542

27. Zovi, D.D.A., Macaulay, S.A.: Attacking automatic wireless network selection. In: Proceedings from the Sixth Annual IEEE SMC Information Assurance Workshop. IEEE, West Point (2005)

28. Mrncciew: 802.11 Mgmt: Beacon Frame (2014). https://mrncciew.com/2014/10/08/802-11-mgmt-beacon-frame/. Accessed 14 Jan 2019

Improvement of the Binary Varshamov Bound

Dejan Spasov[✉]

Faculty of Computer Science and Engineering, Skopje, North Macedonia
dejan.spasov@finki.ukim.mk

Abstract. It has been known that a binary code with parameters $[n, k, d]$ does exist provided that k Hamming balls with diameter d-2 may be fitted in n-dimensional binary field. This simple bound, known as the Varshamov bound, relies on the fact that the number of linear combinations of d vectors from a collection of n vectors cannot be larger than the number of combinations of d or less elements from an n-element set. In this paper, we present several results that extend this counting mechanism and improve the binary Varshamov bound.

Keywords: Linear codes · Greedy codes · Lexicodes · Gilbert-Varshamov bound · Greedy algorithms

1 Introduction

Given an n-dimensional vector space F_q^n over some finite field F_q, a *code C* is any subset of M elements. Let $d(x, y)$ denotes *Hamming distance* between two vectors x and y, i.e. the number of coordinates in which they differ; then we can define *minimum distance d* of a code as $d = \min(d(x, y)), \forall x, y \in C, x \neq y$. We write (n, M, d) to denote a code of M elements and minimum distance d over F_q^n [4].

In this paper we focus on *linear codes*, where the $M = q^k$ codewords form a k-dimensional subspace in F_q^n. We write $[n.k.d]$ to denote a linear code of dimension k and minimum distance d. A linear code C may be defined with a $k \times n$ generator matrix G and $(n - k) \times n$ parity check matrix H, such that $HG^T = 0$. Throughout the paper, we will assume that the generator and parity check matrices are in standard form, namely $G = [I \quad A]$, and $H = [-A^T \quad I]$.

One of the fundamental properties of the parity check matrix H of a linear $[n, k, d]$ code C is that all linear combinations of $d - 1$ columns of H are independent [4]. Relying on this property we can prove existence of new codes from old. Let $H(n, k, d - 2)$ denote the set of all $(n - k)$-touples that are linear combination of $d - 2$ or fewer columns of H. Then C can be extended to a code with parameters $[n + 1, k + 1, d]$ provided

$$|H(n, k, d - 2)| \leq q^{n-k} - 2 \tag{1}$$

In other words, if (1) holds true, then there exists a nonzero vector that is not $d - 2$ linear combination of columns of H. Adding this vector as a column to H does not violate the property of linear independence of all $d - 1$ combinations of columns of H.

© Springer Nature Switzerland AG 2019
S. Gievska and G. Madjarov (Eds.): ICT Innovations 2019, CCIS 1110, pp. 65–71, 2019.
https://doi.org/10.1007/978-3-030-33110-8_6

Let $V(n,d)$ denote the number of combinations of d or less elements from an n-set, namely

$$V(n,d) = \sum_{i=0}^{d} \binom{n}{i} \tag{2}$$

Theorem 1: [Varshamov bound]: A code with parameters $[n, k, d]$ does exist provided

$$V(n-1, d-2) \leq 2^{n-k} - 2 \tag{3}$$

Proof: It is obvious that $|H(n-1, k, d-2)| \leq V(n-1, d-2)$. Thus an $[n-1, k-1, d]$ code can be extended to an $[n, k, d]$ code. Since $V(n-2, d-2) \leq V(n-1, d-2)$ the $[n-1, k-1, d]$ code can be derived from an $[n-2, k-2, d]$ code, and so on.

In the following section we present four theorems that improve the Varshamov bound. The common assumption is that code with parameters $[n, k, d]$ does exist. It is also assumed that neither the generator matrix G nor the parity check matrix H are known. The goal is to extend this code to a code with parameters $[n+l, k+l-1, d]$ and codimension $n-k+1$. Throughout the paper we will write $i+ = 2$ in the index of summation $\sum_{\substack{i=i_0 \\ i+=2}}$ to denote that summation index i increments with step 2 starting from $i = i_0$. The remainder from the division of two integers a and b will be denoted as $a \bmod b$, and we will assume that $\binom{a}{b} = 0$ when $a < b$. Other words, if (1) holds true, then there exists a nonzero vector that is not $d-2$ linear combination of columns of H. Adding this vector as a column to H does not violate the property of linear independence of all $d-1$ combinations of columns of H.

2 Main Results

We will start with a simple improvement of the Varshamov bound.

Theorem 2: *Let the minimum distance d be an even number. Then a code with parameters $[n, k, d]$ does exist provided*

$$V(n-2, d-3) \leq 2^{n-k-1} - 2 \tag{4}$$

Proof: According to (4) and Theorem 1, a code with parameters $[n-1, k, d-1]$ does exist. Adding overall parity check, proves the existence of a code with parameters $[n, k, d]$.

It is obvious that for $d \leq n/2$, Theorem 2 is stronger bound than the Varshamov bound (3). However, it only shows existence of codes with dimension better for at most 1 than the Varshamov bound. For example, the Varshamov bound may show existence of a code with parameters $[n, k, d]$, while (4) may show existence of a code with parameters $[n, k+1, d]$.

The following theorem was first published in [6], but with an unnecessary condition that specifies the parity check matrix H of the $[n, k, d]$ code be given.

Theorem 3: *Any $[n, k, d]$ code can be extended to a code with parameters $[n+l, k+l-1, d]$ provided*

$$\sum_{\substack{i=1 \\ i+=2}}^{d-2} \binom{l-1}{i} V(n, d-2-i) \leq 2^{n-k} - 1 \tag{5}$$

Proof: Let $m = n - k$. We build the parity check matrix H_{m+1} of the $[n+l, k+l-1, d]$ code from the parity check matrix H_m of the $[n, k, d]$ code by greedily adding columns to $\begin{bmatrix} 0 \cdots 0 \\ H_m \end{bmatrix}$ with first row equal to 1. Hence, the resulting matrix H_{m+1} will have the form

$$H_{m+1} = \begin{bmatrix} 0 \cdots 0 & 1 \cdots 1 \\ H_m & L_m \end{bmatrix} \tag{6}$$

Let l be the number of columns of $\begin{bmatrix} 1 \cdots 1 \\ L_m \end{bmatrix}$. We are interested in estimating the number of $(d-2)$-linear combinations spanned by the columns of H_{m+1} whose first row is 1. If this number is less than 2^{n-k} we can keep adding columns to $\begin{bmatrix} 1 \cdots 1 \\ L_m \end{bmatrix}$.

Each of the combinations must include odd number of vectors i from $\begin{bmatrix} 1 \cdots 1 \\ L_m \end{bmatrix}$ and at most $d - 2 - i$ vectors from $\begin{bmatrix} 0 \cdots 0 \\ H_m \end{bmatrix}$. We can choose i columns from $\begin{bmatrix} 1 \cdots 1 \\ L_m \end{bmatrix}$ in $\binom{l}{i}$ ways and at most $d - 2 - i$ columns from $\begin{bmatrix} 0 \cdots 0 \\ H_m \end{bmatrix}$ in $V(n, d-2-i)$ ways. Since L_m and H_m are two disjoint sets, for fixed i we need the product of the two quantities $\binom{l}{i} V(n, d-2-i)$. Thus, the number of combinations from H_{m+1} whose first row equals to 1 is sum over the odd values of i, namely we conclude (5).

To obtain code parameters that satisfy (5), first we must prove existence of a code with parameters $[n, k, d]$. To show the existence of an $[n, k, d]$ code we can use the Varshamov bound (3), thus solving one additional inequality, or we can use (5) in recurrent fashion. With the recurrent method, we use (5) to prove existence of an $[n, k, d]$ code. Then we use (5) again for the $[n+l, k+l-1, d]$ code. The repetition $[d, 1, d]$ code can be used as stopping criterion for the recurrent method. It is expected that the latter method gives better results but requires $n - k - d$ inequalities to be solved. We should note that in each recurrent step $l \geq 1$, thus providing the *trivial lengthening* (as defined in [1]). Forcing $l \leq 2$ we obtain Elia's result [3].

Next, we will assume that additional information is known about the parity check matrix of the initial code – the number of columns of H with nonzero first row. We will use this information to improve Theorem 3.

Theorem 4: *Let assume that a code with parameters $[n, k, d]$ does exist. Let l_{m-1} be the number of columns of its parity check matrix whose first row is 1. Then there exist a code with parameters $[n + l_m, k + l_m - 1, d]$, where $l_m = l'_m + l''_m$, provided that l'_m, l''_m satisfy*

$$\sum_{\substack{i=1 \\ i+=2}}^{d-2} \binom{l'_m-1}{i} \sum_{\substack{j=0 \\ j+=2}}^{d-2-i} \binom{l_{m-1}}{j} V(n - l_{m-1}, d - 2 - i - j) \leq 2^{n-k-1} - 1 \qquad (7)$$

$$\sum_{\substack{i=0 \\ i+=2}}^{d-2} \binom{l''_m-1}{i} \sum_{\substack{j=(i+1)\bmod 2 \\ j+=2}}^{d-2-i} \binom{l'_m}{j} \sum_{\substack{k=(i+1)\bmod 2 \\ k+=2}}^{d-2-i-j} \binom{l_{m-1}}{k} V(n - l_{m-1}, d - 2 - i - j - k) \leq 2^{n-k-1} - 1$$

$$(8)$$

Proof: We will assume that $H_m = \begin{bmatrix} 0 \cdots 0 & 1 \cdots 1 \\ H_{m-1} & L_{m-1} \end{bmatrix}$ is the parity check matrix of the $[n, k, d]$ code. Starting from H_m we greedily build the parity check matrix H_{m+1} by adding vectors from F_2^{m+1}. Eventually, H_{m+1} will have the following form

$$H_m = \begin{bmatrix} 0 \cdots & \cdots 0 & 1 \cdots & \cdots 1 \\ 0 \cdots 0 & 1 \cdots 1 & 0 \cdots 0 & 1 \cdots 1 \\ H_{m-1} & L_{m-1} & L'_m & L''_m \end{bmatrix}$$

Let n_{m-1}, l_{m-1}, l'_m, and l''_m be the number of columns of H_{m-1}, L_{m-1}, L'_m, L''_m, respectively. In similar fashion as in Theorem 3 we are interested in the number of $d - 2$ or less combinations of columns of H_{m+1} that result in a vector of the form $\begin{bmatrix} 1 & x_1 & x_2 & x_3 & \cdots & x_{m+1} \end{bmatrix}^T$. Each combination must include odd number of vectors $i, i \leq d - 2$, from $\begin{bmatrix} 1 \cdots 1 \\ 0 \cdots 0 \\ L'_m \end{bmatrix}$, even number $j, j \leq d - 2 - i$, vectors from $\begin{bmatrix} 0 \cdots 0 \\ 1 \cdots 1 \\ L_{m-1} \end{bmatrix}$, and at most $d - 2 - i - j$ vectors from $\begin{bmatrix} 0 \cdots 0 \\ 0 \cdots 0 \\ H_{m-1} \end{bmatrix}$. Since these matrices are disjoint set of vectors, the number of $d - 2$ or less combinations from H_{m+1} that result in a $\begin{bmatrix} 1 & 0 & x_2 & x_3 & \cdots & x_{m+1} \end{bmatrix}^T$ vector is

$$\sum_{\substack{i=1 \\ i+=2}}^{d-2} \binom{l'_m-1}{i} \sum_{\substack{j=0 \\ j+=2}}^{d-2-i} \binom{l_{m-1}}{j} V(n - l_{m-1}, d - 2 - i - j)$$

This proves (7). Next, we are interested in the number of $d-2$ or less combinations of columns of H_{m+1} that result in a vector of the form $[1 \quad 1 \quad x_2 \quad x_3 \quad \cdots \quad x_{m+1}]^T$. We will consider two cases:

1. A combination of even number i of columns from $\begin{bmatrix} 1 \cdots 1 \\ 1 \cdots 1 \\ L_m'' \end{bmatrix}$ must have odd number

$j, j \le d-2-i$ columns from $\begin{bmatrix} 1 \cdots 1 \\ 0 \cdots 0 \\ L_m' \end{bmatrix}$ and odd number $k, k \le d-2-i-j$, of

columns from $\begin{bmatrix} 0 \cdots 0 \\ 1 \cdots 1 \\ L_{m-1} \end{bmatrix}$. Then we can add $V(n_m, d-2-i-j-k)$ columns from

$\begin{bmatrix} 0 \cdots 0 \\ 0 \cdots 0 \\ H_{m-1} \end{bmatrix}$. Thus, the number of $d-2$ or less combinations from H_{m+1} that result

in a $[1 \quad 1 \quad x_2 \quad x_3 \quad \cdots \quad x_{m+1}]^T$ vector when i is even is

$$\sum_{\substack{i=0 \\ i+=2}}^{d-2} \binom{l_m''-1}{i} \sum_{\substack{j=1 \\ j+=2}}^{d-2-i} \binom{l_m'}{j} \sum_{\substack{k=1 \\ k+=2}}^{d-2-i-j} \binom{l_{m-1}}{k} V(n_{m-1}, d-2-i-j-k) \qquad (9)$$

2. A combination of odd number i of columns from $\begin{bmatrix} 1 \cdots 1 \\ 1 \cdots 1 \\ L_m'' \end{bmatrix}$ must have even number

$j, j \le d-2-i$, columns from $\begin{bmatrix} 1 \cdots 1 \\ 0 \cdots 0 \\ L_m' \end{bmatrix}$ and even number $k, k \le d-2-i-j$,

columns from $\begin{bmatrix} 0 \cdots 0 \\ 1 \cdots 1 \\ L_{m-1} \end{bmatrix}$. In addition, we can add $V(n_m, d-2-i-j-k)$ columns

from $\begin{bmatrix} 0 \cdots 0 \\ 0 \cdots 0 \\ H_{m-1} \end{bmatrix}$. Hence the number of $d-2$ or less combinations from H_{m+1} that

result in a $[1 \quad 1 \quad x_2 \quad x_3 \quad \cdots \quad x_{m+1}]^T$ vector, when i is odd is equal to

$$\sum_{\substack{i=1 \\ i+=2}}^{d-2} \binom{l_m''-1}{i} \sum_{\substack{j=0 \\ j+=2}}^{d-2-i} \binom{l_m'}{j} \sum_{\substack{k=0 \\ k+=2}}^{d-2-i-j} \binom{l_{m-1}}{k} V(n_{m-1}, d-2-i-j-k) \qquad (10)$$

The left-hand side of (8) is sum of (9) and (10).

Though Theorem 4 looks complex it can be solved recurrently, i.e. first we use (7) and (8) to secure existence of a $[n, k, d]$ code with codimension m-1, and then we use it again and prove the existence of a code with parameters $[n + l_m, k + l_m - 1, d]$ and codimension m. We should mention that $l'_m \geq 1$ thus allowing the trivial lengthening, but $l''_m \geq 0$.

3 Conclusion

We have shown several results that improve the estimate of the number of d linear combinations from a set of k vectors (1) and thus improve the Varshamov bound. The Varshamov bound considers the parity check matrix H as set with n elements, wherein columns of the parity check matrix are considered elements of the set. A linear combination of columns of H is considered a combination of elements of the set. The problem with this counting mechanism is that two linear combinations may evaluate to the same vector, but they will be counted twice on the left-hand side of (3). Theorem 3 improves the Varshamov bound by developing a counting mechanism where certain linear combinations that evaluate to the same vector are counted only once. Theorem 4 improves Theorem 3 and the Varshamov bound by developing more sophisticated counting mechanism where more linear combinations that evaluate to the same vector are counted only once. A question that naturally arises in this context is if we can further improve Theorem 4 by using additional assumptions for the structure of the parity check matrix of the $[n, k, d]$ code. For example, we can include the second row of H in the considerations and assume that the number of columns that start with $[0 \ 0 \ \cdots]^T$, $[0 \ 1 \ \cdots]^T$, $[1 \ 0 \ \cdots]^T$, and $[1 \ 1 \ \cdots]^T$ is known. Continuing along these lines, we will obtain a complex system of four inequalities that will improve (7) and (8). But with each additional structural assumption about H, the number of inequalities doubles, thus ultimately leading to unsolvable system of 2^m inequalities.

Jiang and Vardy [5], (later improved in [8]) have asymptotically improved the Gilbert bound by showing that there exist a nonlinear (n, M, d) code such that

$$M \geq c \cdot \frac{2^n}{V(n, d - 1)} \cdot n \tag{11}$$

In [2], it has been proved that linear binary $[n, \frac{n}{2}]$ random double-circulant codes satisfy the Jiang-Vardy bound. One may ask how Theorem 4 compares with the Jiang-Vardy bound. We leave this question as open, but we believe that (7) and (8) are better bound than (11) in terms of proving better code parameters. Our belief is based on the comparison between the Varshamov bound (3) and the Jiang-Vardy bound (11). Comparing these bounds, we can observe that the Jiang-Vardy bound is better that the Varshamov bound for a factor n only when $\frac{d}{n} \to 0$. However, at the other side of the spectrum, where $\frac{d}{n} \to const$, these bounds differ by a constant.

References

1. Barg, A., Guritman, S., Simonis, J.: Strengthening the Gilbert-Varshamov bound. Linear Algebra Appl. **307**(1–3), 119–129 (2000)
2. Gaborit, P., Zemor, G.: Asymptotic improvement of the Gilbert-Varshamov bound for linear codes. IEEE Trans. Inf. Theory **54**(9), 3865–3872 (2008)
3. Elia, M.: Some results on the existence of binary linear codes (Corresp.). IEEE Trans. Inf. Theory **29**(6), 933–934 (1983)
4. MacWilliams, F.J., Sloane, N.J.A.: The Theory of Error-Correcting Codes, vol. 16. Elsevier (1977)
5. Jiang, T., Vardy, A.: Asymptotic improvement of the Gilbert-Varshamov bound on the size of binary codes. IEEE Trans. Inf. Theory **50**(8), 1655–1664 (2004)
6. O'Brien, K.M., Fitzpatrick, P.: Bounds on codes derived by counting components in Varshamov graphs. Des. Codes Crypt. **39**(3), 387–396 (2006)
7. Tolhuizen, L.M.: The generalized Gilbert-Varshamov bound is implied by Turan's theorem [code construction]. IEEE Trans. Inf. Theory **43**(5), 1605–1606 (1997)
8. Vu, V., Wu, L.: Improving the Gilbert-Varshamov bound for q-ary codes. IEEE Trans. Inf. Theory **51**(9), 3200–3208 (2005)

Electrical Energy Consumption Prediction Using Machine Learning

Simon Stankoski[1], Ivana Kiprijanovska[1], Igor Ilievski[2],
Jovanovski Slobodan[2], and Hristijan Gjoreski[1(✉)]

[1] Faculty of Electrical Engineering and Information Technologies,
University of Ss. Cyril and Methodius in Skopje, Skopje, North Macedonia
`simonstankoski997@gmail.com`,
`kiprijanovskai@gmail.com`, `hristijang@feit.ukim.edu.mk`
[2] ITS Iskratel, Skopje, North Macedonia
`ilievski@its-sk.com.mk`, `jovanovski@iskratel.si`

Abstract. The paper presents a Machine Learning (ML) approach to household Electrical Energy (EE) consumption prediction. It includes: data preprocessing, feature engineering, learning a classification model, and experimental evaluation on one of the largest datasets for household EE consumption – DataPort dataset. Beside the features extracted on the historical EE consumption, we additionally analyze weather and contextual-calendar related features. We believe that the combination of multiple sources of data (calendar, weather, historical EE consumption) provides more information to the model in order to learn better performing model. The experimental results showed that in all the cases the ML algorithms outperform the baselines, with the best performing the XGBoost - achieved 0.69 RMSE score, 0.41 MAE score and 0.67 R^2 score which is significantly better than the best performing baseline model (the value from 24 h ago). Additionally, the results show that the largest errors are made for the weekends, which was expected due to the irregularities in the schedule - trips, vacations, etc.

Keywords: Energy consumption · Prediction · Forecast · Machine learning · Deep learning · Household · Day-ahead

1 Introduction

Electrical energy (EE) consumption is essential in our daily life. In recent years, with the concept of deregulation in the power industry, many challenges have been faced by the participants in the electricity market. The excessive produce of electrical energy is a problem because the storage of excess electricity is difficult and challenging. Therefore, a system that can accurately predict the EE consumption can significantly reduce the problems with overproduction and storage of the EE. Such system can help to optimize the production and the consumption of EE. Also, it may potentially decrease the EE consumption costs for each individual household by better planning the production/ buying the EE in advance.

© Springer Nature Switzerland AG 2019
S. Gievska and G. Madjarov (Eds.): ICT Innovations 2019, CCIS 1110, pp. 72–82, 2019.
https://doi.org/10.1007/978-3-030-33110-8_7

The massive deployment of smart grid technologies in the residential sector brings both opportunities and increased challenges to the load forecasting community. The electricity consumption can be accessed in close to real time and allows both the demand and supply side to extract valuable information for efficient management of the electrical network load.

In this paper we present a machine learning approach to household EE consumption prediction. The approach includes: data preprocessing, feature engineering, learning a classification model, and experimental evaluation on one of the largest datasets for household EE consumption.

2 Related Work

Most of the approaches for prediction of the consumption of the electrical energy can be classified as conventional statistical methods, machine learning (ML) methods and deep learning (DL) methods. Statistical approaches usually use mathematical function to model the load with several input factors. These methods include curve fitting, data extrapolation and smoothing techniques, and are usually based on the assumption that the load data have an internal structure. Some examples of statistical based approaches in the field of short-term load forecasting (STLF) in the literature are: multiple regression method [15], exponential smoothing [1, 12], adoptive load forecasting [21], and stochastic time series [13, 16, 20]. However, the load forecasting is a complex multi-variable and multi-dimensional estimation problem and these approaches are not always able to find the implicit nonlinear relationship between the load and the influencing variables. On the other hand, ML methods are able to find some regularities and patterns in the historical data and use them to forecast the future load consumption. In the literature, the most common ML algorithms used for STLF are: Support Vector Machines [17], Random Forest [4, 8] and Artificial Neural Networks (ANN) [11, 12]. In recent years, DL has been a subject undergoing intense study in many fields, especially in load forecasting. DNN have demonstrated capability to do non-linear curve-fitting and approximate any complex functions with arbitrary precision. The most popular artificial neural network architecture for electric load forecasting is back propagation, as presented in [19, 22, 24].

Extensive and comprehensive review papers on point load forecasting at aggregated level already exist. However, the literature on individual household load forecasting is limited. That is because it is widely acknowledged that short-term load forecasting (STLF) at such granular level is extremely challenging due to significant uncertainty and volatility underlying the smart metering data. At this level, uncertainty is more influenced by customer behavior, which is too stochastic to predict. Therefore, the nature of our challenge is to forecast load with significant uncertainty.

In general, most of the papers present satisfactory results within their research, but without comparing the accuracy of the used ML methods to standard benchmarks and existing methods, which is essential in order to prove that the proposed methods introduces some improvements in the load forecasting field, as stated in [14]. Our proposed solution in the energy consumption forecast includes a short-term forecast (day ahead hourly forecasts) for consumer-level consumption (a special forecast for

each hour of the next day), which has great industrial and economic value. Except historical electrical load, we take into account other factors that can affect the actual energy consumption of a residential house, such as weather conditions (i.e. in winter and summer, a large percent of the electrical power is consumed for heating and cooling), and the type of day (working day, holiday).

3   Dataset and Preprocessing

In this research we have studied multiple datasets for EE consumption. After the analysis we chose the Dataport dataset [6], which is one of the largest sources of detailed, disaggregated customer energy data. The database contains electricity data collected from more than 1000 households in the US, mainly in Austin, Texas. The dataset stores the actual measurements of every circuit in these houses, collected by eGauge devices [9]. Additionally, it contains extensive weather data for the observed region. Short-term load forecasting is mainly affected by weather parameters, because heating, ventilation, and air conditioning (HVAC) are strongly dependent on the outdoor temperature, wind speed, humidity, etc.

In the next phase we performed preprocessing of the dataset, including:

- Dealing with instances that contain incorrect values for some of the columns – We came across some negative values for consumed electricity. In this case, we provided the required value based on values from other columns. We calculated the total load consumption as sum of the power drawn from the electrical grid and consumed power from the solar system grid.
- Dealing with instances that have column values that deviate from the expected range (outlier values) or lack values for some of the columns – If the incorrect value referred to some of the weather related variables, we simply found the true value among the other instances referring to the same hour of the same day. However, if the value of the load consumption was incorrect even after the first step, we deleted the instance.
- Dealing with duplicate values for some of the columns - There were some cases when the measuring devices set in particular households gave the same value over a longer period of time. This led to a conclusion that there had to be a problem with the sensor itself. Since we had a dataset and a large number of households included, we were able to delete all instances referring to those households without substantial loss of data.

After these steps, the filtered dataset was ready for further processing. The average load consumption per household over the 2016 and 2017 can be observed in Fig. 1.

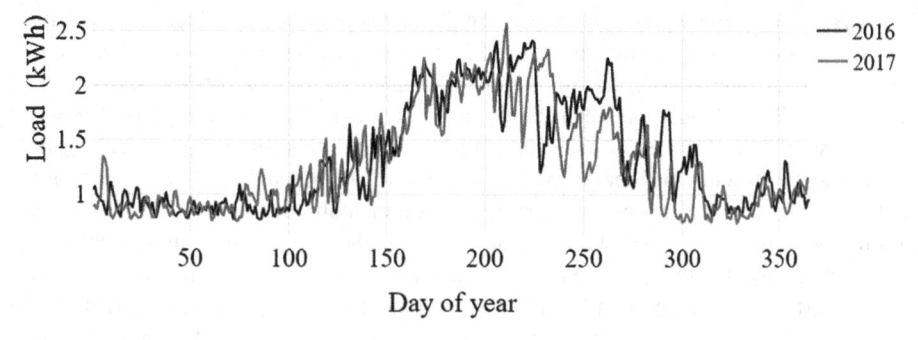

Fig. 1. Average EE load consumption per household for the 2016 and 2017.

4 Methods

4.1 Feature Extraction

The electrical load consumption is a complex time series driven by multiple factors. After a thorough analysis, we extracted different types of features: calendar, weather, interaction and historical load.

The calendar features include seasons, days, calendar data, holidays, etc.

The weather features include temperature, humidity, precipitation intensity and wind speed, etc. Since the effect of outdoor temperature on load consumption is obviously significant, we included more temperature related features, such as temp_beforeX, which represents the temperature of the previous X-th hour, and the daily average temperature from two days ago.

Interaction features represent the features that are calculated by interaction with another feature. The hours in different days of a week may result in different load due to human activities. For instance, there may be lower load in the morning of the weekends than the other mornings, because people do not have to get up as early as weekdays to go to work, which results in less load consumption at home. We implemented the interaction by multiplying two variables.

It is also very intuitive that load consumption is highly related to historic loads. We included the load consumption values by a single user at the same hour of some of the previous days, as well as the average values of consumption from all the users included at the same hours. Based on these previously extracted features, we generated more statistical input data (for example: standard deviation, median, variance…) out of four different types of time series, that give new insights into the time series dynamics.

In total we have calculated 96 features.

4.2 Machine Learning Algorithms

We have used the following regression learning algorithms:

Linear Regression [7] is a statistical method that allows us to study the relationship between variables. It attempts to model the relationship by fitting a linear equation to observed data. It works by minimizing the total of the square of the errors.

K-Nearest Neighbors (kNN) [5] is a simple algorithm that uses 'feature similarity' to forecast values of any new data points. This means that the new point is assigned a value by local interpolation of the targets associated of the nearest neighbors in the training set. The distance between the points is measured by Manhattan distance.

Decision Tree Regressor [25] is an algorithm that builds regression models in the form of a tree structure. It breaks down a dataset into smaller and smaller subsets while at the same time an associated decision tree is incrementally developed. The final result is a tree with decision nodes with two or more branches, each representing values for the attribute tested, and leaf nodes which represent a decision on the numerical target.

Support Vector Machines [23] are very specific class of algorithms, characterized by usage of kernels, absence of local minima, sparseness of the solution and capacity control obtained by acting on the margin, or on number of support vectors, etc.

Gradient boosting [10, 11] is a ML technique for regression which produces a forecast model in the form of an ensemble of weak forecast models, typically decision trees. It builds the model in a stage-wise fashion like other boosting methods do, and it generalizes them by allowing optimization of an arbitrary differentiable loss function.

XGBoost [3] is a highly sophisticated gradient boosting algorithm. It has implemented regularization which reduces over fitting and in-built routine that handles missing values. As well as that, it implements parallel processing which makes it much faster than other gradient algorithms.

5   Evaluation Setup

We used Root Mean Squared Error (RMSE), Mean Absolute Error (MAE) and R^2 score to evaluate the methods. They are defined as follows:

$$RMSE = \sqrt{\frac{1}{n}\sum_{i=1}^{n}\left(|True - \Pr edicted|\right)^2} \qquad (1)$$

$$MAE = \frac{1}{n}\sum_{i=1}^{n}|True - \Pr edicted| \qquad (2)$$

$$R^2 = 1 - \frac{\sum_{i=1}^{n}(True - \Pr edicted)^2}{\sum_{i=1}^{n}(True - Average)^2} \qquad (3)$$

where n is the number of data samples. Both MAE and RMSE express average model forecast error in units of the variable of interest – kilowatt hour (kWh). Both can range from 0 to infinity and are indifferent to the direction of errors. They are negatively – oriented scores, which means lower values are better. R^2 score on the contrary, is positively - oriented score, where the best possible value is 1. It provides a measure of how well observed outcomes are replicated by the model.

The data is collected from 925 households, for a period of 3 years (2015, 2016, and 2017) and partially 2018 (10 months). We sliced the data to three pieces: the models were trained on a total of 27 months of measured hourly load data, 6 months were taken for validation (model selection and parameter optimization), and a blind year of test data was withheld until the final testing of the forecasting capabilities of the models.

6 Experimental Results

Figures 2, 3 and 4 show the RMSE, the MAE and the R^2 score for each of the algorithms respectively. Additionally, we included different baselines for comparison:

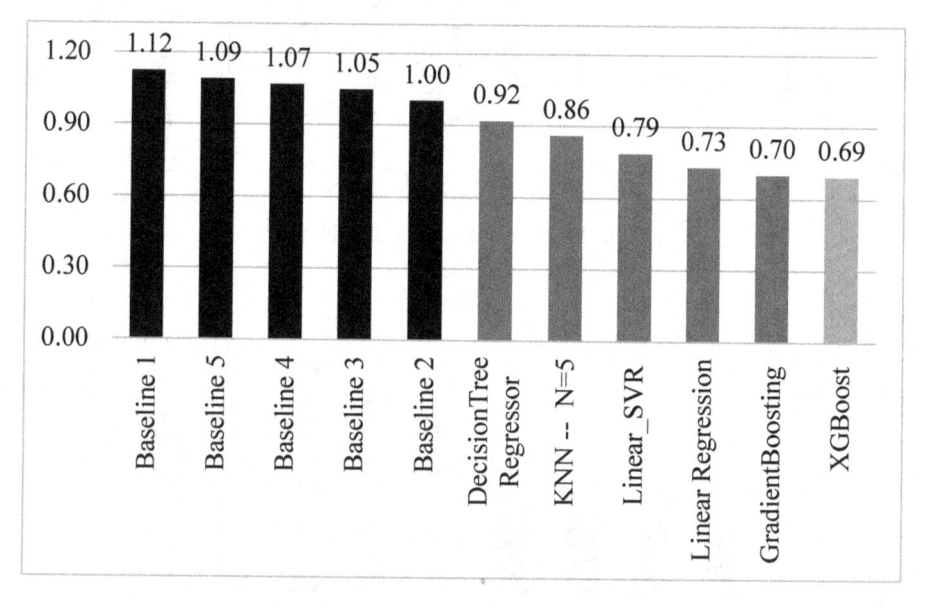

Fig. 2. RMSE comparison of algorithms.

– Baseline 1 - provides the average amount of used EE from all the instances in the training set.
– Baseline 2 - provides the amount of used EE by the specific user 24 h before the prediction hour.
– Baseline 3 - provides the EE used by the specific user 168 h (1 week) before the hour of prediction.
– Baseline 4 - provides the average EE used 24 h before the prediction hour.
– Baseline 5 - provides the average amount of used EE 168 h (1 week) before the hour that we want to predict.

The results show that in all the cases the ML algorithms outperform the Baselines, with the best performing the Gradient Boost and the XGBoost. The XGBoost achieved 0.69 RMSE score, 0.41 MAE score and 0.67 R^2 score which is significantly better than the best performing Baseline model (last value - from 24 h ago).

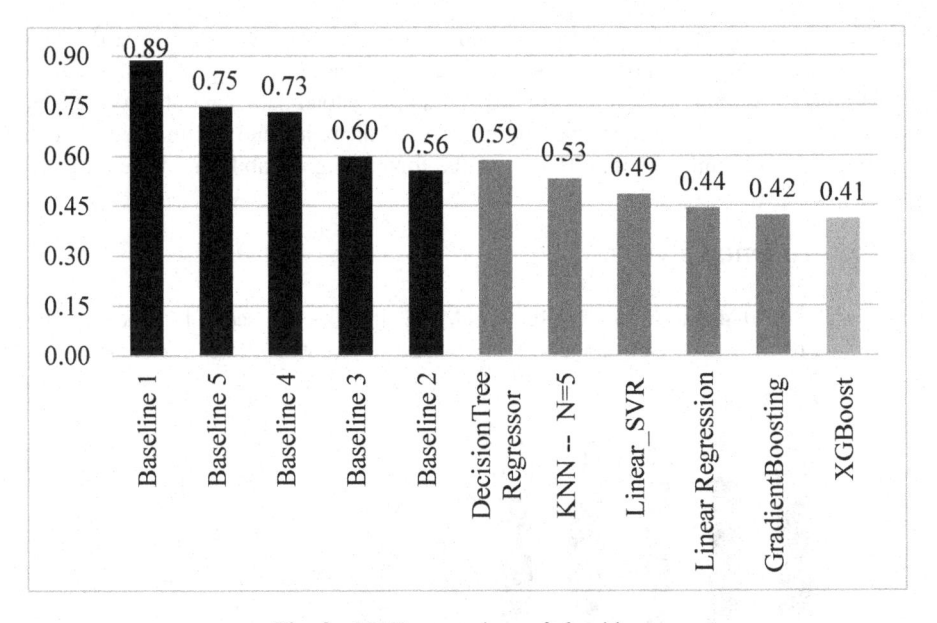

Fig. 3. MAE comparison of algorithms.

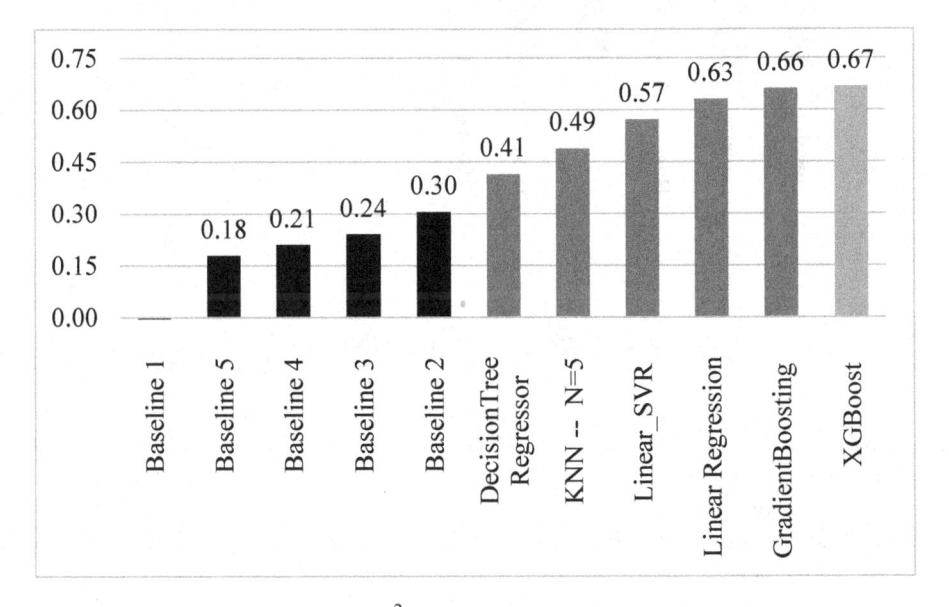

Fig. 4. R^2 comparison of algorithms

Figures 4, 5 and 6 show the RMSE, MAE and R^2 score achieved for each of the days of the week. The results are obtained by averaging the errors for all the users for each day. The results show the ML algorithms significantly outperform the Baseline. Additionally, the results show that the largest errors are made for the weekends. This

was kind of expected, because the weekends are more challenging to forecast due to trips, vacations and irregularities in our everyday life. This suggests using separate models for the weekends (Fig. 7).

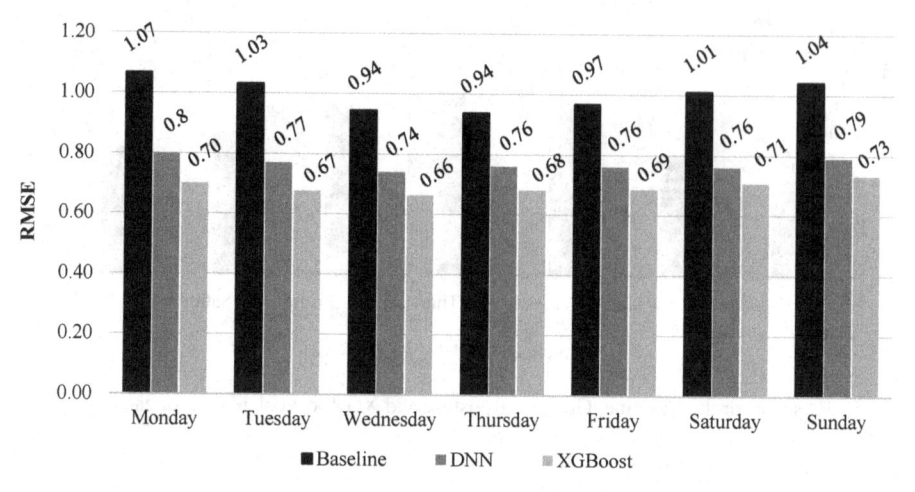

Fig. 5. RMSE for the Baseline, Gradient Boosting and XGBoost calculated for each day of the week.

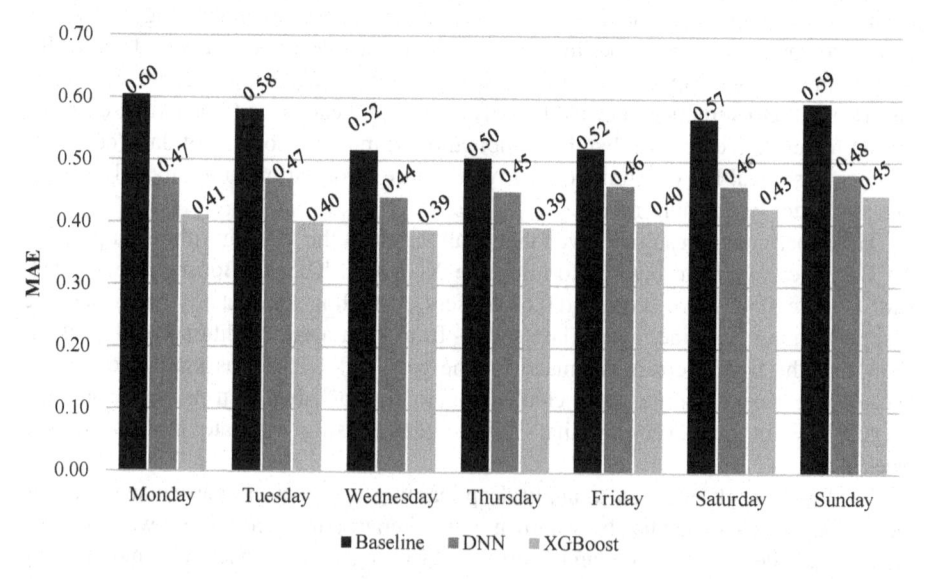

Fig. 6. MAE for the Baseline, Gradient Boosting and XGBoost calculated for each day of the week.

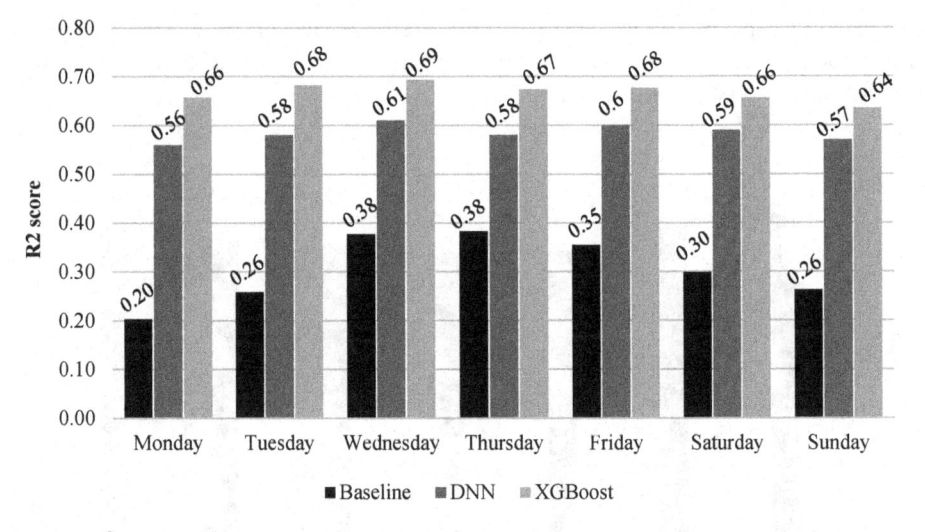

Fig. 7. R^2 score for the Baseline, Gradient Boosting and XGBoost calculated for each day of the week

7   Conclusion

The paper presented a machine learning approach to household EE consumption prediction. The approach includes: data preprocessing, feature engineering, learning a classification model, and experimental evaluation on one of the largest datasets for household EE consumption – DataPort dataset. Beside the features extracted on the historical EE consumption, we additionally extracted weather and contextual-calendar related features. We believe that the combination of multiple sources of data (calendar, weather, historical EE consumption) provides more information to the model in order to learn better performing model.

The experimental results showed that in all the cases the ML algorithms outperform the Baselines, with the best performing the XGBoost. The XGBoost achieved 0.69 RMSE score, 0.41 MAE score and 0.67 R^2 score, which is significantly better than the best performing Baseline model (last value - from 24 h ago). Additionally, the results show that the largest errors are made for the weekends. This was kind of expected, because the weekends are more challenging to forecast due to trips, vacations and irregularities in our everyday life. This suggests using separate models for the weekends.

For future work, we plan to apply deep learning approach, in particular Recurrent Neural Network – which has been shown in the literature to perform very well on time-series data. Additionally, we plan to perform clustering of the households, and this way to have different prediction models for different clusters of users. We believe this will be beneficial and will improve the performance, because there are different types of users (elderly users that are mostly home, active users that are mostly outside working, etc.) and it is more challenging to the model to try to learn a model for all these types of users.

Acknowledgment. We gratefully acknowledge the support of NVIDIA Corporation with the donation of the Titan Xp GPU used for this research. Additionally, we would like to thank the DataPort portal for providing the dataset for research purposes.

References

1. Adilah, N., Jalil, A., Ahmad, M.H., Mohamed, N.: Electricity load demand forecasting using exponential smoothing methods. World Appl. Sci. J. **22**(11), 1540–1543 (2013)
2. Bakirtzis, A.: A neural network short term load forecasting model for the greek power system. IEEE Trans. Power Syst. **11**(2), 858–863 (1996)
3. Chen, T.: XGBoost: a scalable tree boosting system. ArXiv e-prints (2016)
4. Cheng, Y.-Y., Chan, P., Qiu, Z.-W.: Random forest based ensemble system for short term load forecasting. In: 2012 International Conference on Machine Learning and Cybernetics (ICMLC), Xian, vol. 1, pp. 52–56 (2012)
5. Aha, D.: Instance-based learning algorithms. Mach. Learn. **6**(1), 37–66 (1991)
6. Dataport dataset used for this research. https://www.pecanstreet.org/dataport/about/
7. Freedman, D.A.: Statistical Models: Theory and Practice, p. 26. Cambridge University Press, Cambridge (2009). https://doi.org/10.1007/3-540-62858-4, https://doi.org/10.1109/smartgrid comm.2018.8587489
8. Dudek, G.: Short-term load forecasting using random forests. In: Filev, D., Jablkowski, J., Kacprzyk, J., Krawczak, M., Popchev, I., Rutkowski, L., et al. (eds.) Intelligent Systems'2014. AISC, vol. 323, pp. 821–828. Springer, Cham (2015). https://doi.org/10.1007/978-3-319-11310-4_71
9. Egauge device. https://www.egauge.net/
10. Friedman, J.H.: Greedy function approximation: a gradient boosting machine. Ann. Stat. **29**(5), 1189–1232 (2001)
11. Friedman, J.H.: Stochastic gradient boosting. Comput. Stat. Data Anal. **38**(4), 367–368 (2002)
12. Göb, R.: Electrical load forecasting by exponential smoothing with covariates. Appl. Stoch. Models Bus. Ind. **29**(6), 629–645 (2013). http://dx.doi.org/10.1109/ICMLC.2012.6358885, https://www.egauge.net/. Accessed 28 May 2019, https://www.pecanstreet.org/dataport/. Accessed 15 May 2019
13. Huang, S.-J.: Short-term load forecasting via ARMA model identification including non-Gaussian process considerations. IEEE Trans. Power Syst. **18**(2), 673–679 (2003)
14. International Journal of Forecasting, Official Publication of the International Institute of Forecasters. https://robjhyndman.com/hyndsight/benchmarks/?fbclid=IwAR0o34h8CSmk KGpqNdlFRHvSj_ib5bNL5MihUQLXSYyLKtdD6xTbOhfw_tA. Accessed 10 June 2019
15. Kim, J., Cho, S., Ko, K., Rao, R.R.: Short-term electric load prediction using multiple linear regression method. In: 2018 IEEE International Conference on Communications, Control, and Computing Technologies for Smart Grids (SmartGridComm), Aalborg, pp. 1–6 (2018)
16. Lee, Y.-S.: Forecasting time series using a methodology based on autoregressive integrated moving average and genetic programming. Knowl.-Based Syst. **24**, 66–72 (2011)
17. Li, G., Cheng, C., Lin, J., Zeng, Y.: Short-term load forecasting using support vector machine with SCE-UA algorithm. In: Third International Conference on Natural Computation (ICNC 2007), Haikou, pp. 290–294 (2007). https://doi.org/10.1109/icnc.2007.660
18. Mandal, P.: Neural networks approach to forecast several hours ahead electricity prices and loads in deregulated market. Energy Convers. Manage. **47**(15–16), 2128–2142 (2006)

19. Marino, D.L., Amarasinghe, K., Manic, M.: Building energy load forecasting using Deep Neural Networks. In: IECON 2016 - 42nd Annual Conference of the IEEE Industrial Electronics Society, Florence, pp. 7046–7051 (2016). https://doi.org/10.1109/iecon.2016.7793413

20. Pappas, S.S.: Electricity demand loads modeling using AutoRegressive Moving Average (ARMA) models. Energy **33**, 1353–1360 (2008)

21. Lu, Q.: An adaptive nonlinear predictor with orthogonal escalator structure for short-term load forecasting. IEEE Trans. Power Syst. **4**(1), 158–164 (1989)

22. Reddy, S.S., Momoh, J.A.: Short term electrical load forecasting using back propagation neural networks. In: 2014 North American Power Symposium (NAPS), Pullman, WA, pp. 1–6 (2014). https://doi.org/10.1109/naps.2014.6965453

23. Shevade, S.K.: Improvements to the SMO algorithm for SVM regression. IEEE Trans. Neural Networks **11**(5), 1188–1193 (2000)

24. Shi, H.: Deep learning for household load forecasting – a novel pooling deep RNN. IEEE Trans. Smart Grid **9**(5), 5271–5280 (2018)

25. Wang, Y., Witten, I.H.: Induction of model trees for predicting continuous classes. In: van Someren, M., Widmer, G. (eds.) Poster papers of the 9th European Conference on Machine Learning, Prague (1997)

Diatom Ecological Modelling with Weighted Pattern Tree Algorithm by Using Polygonal and Gaussian Membership Functions

Andreja Naumoski[✉], Georgina Mirceva, and Kosta Mitreski

Faculty of Computer Science and Engineering, Ss. Cyril and Methodius
University in Skopje, Skopje, North Macedonia
{andreja.naumoski,georgina.mirceva,
kosta.mitreski}@finki.ukim.mk

Abstract. Weighted Pattern Tree (WPT) algorithm is as an extension of the Pattern Tree (PT) algorithm, which could be used for fuzzy modelling. This algorithm utilizes the similarity between two fuzzy sets in order to quantify how much a particular tree model is confident to predict a given class. The Membership Functions (MFs) play an important role in model induction and thus on the model's performance. Therefore, this paper aims to investigate the influence of different MFs, not only by analyzing different mathematical distributions, but also to investigate the influence of the number of MFs per attribute used for fuzzification of the datasets, as well as the different settings of the algorithm in the area of diatom ecological modelling. The experimental results show that WPTs with depth 10 using polygonal MFs with high number of MFs per attribute are excellent for describing the training data, while the models that are built with low number of MFs are excellent for making predictions for unseen data. The results from this research can be used for ecological modelling of diatoms, to classify a given diatom into a particular water quality class.

Keywords: Weighted Pattern Trees · Diatoms · Membership Functions · Statistical significance

1 Introduction

In many water related studies, the aquatic biologists try to find different ways to understand and reveal the relationship between the environment and how that environment influences on the aquatic organism's life cycle. Typically, a water quality class is defined in a certain range of the environmental stress factors in which the organism belongs. In such studies, the ecological experts usually relay on the data that is collected over period of time, and then they apply statistical methods to reveal any hidden pattern that can be found in the data. Usually classical statistical approaches such as canonical correspondence analysis, detrended correspondence analysis and principal component analysis are used as modelling techniques in this research area [13]. Although these techniques provide useful insights in the data, they are limited in terms of interpretability, and in most cases they are influenced by the subjective opinion of the expert.

© Springer Nature Switzerland AG 2019
S. Gievska and G. Madjarov (Eds.): ICT Innovations 2019, CCIS 1110, pp. 83–95, 2019.
https://doi.org/10.1007/978-3-030-33110-8_8

With the advancements in computer science, particularly in the area of machine learning and data mining, where algorithms gain new knowledge from the data at hand, the algorithms provide better understanding of the data compared to the traditional statistical approaches. Many machine learning algorithms have provided better inside look in the data. They also provide models with better interpretability, and at the same time with improved accuracy. One such group of algorithms are decision trees [12], which partition the space in order to find the relationship between the input and output attributes. This could be done using various metrics to measure the benefit of making a given split, which has impact on the algorithm's performance. Also, there is a sub-class of decision tree algorithms that are based on the concepts of fuzzy theory, which are introduced to further improve the performance.

There is an extensive research effort in developing algorithms based on fuzzy set-theory, and many of these advancements are inspired by crisp decision trees [12]. For example, Yuan and Shaw [20] proposed fuzzy decision trees induction using fuzzy entropy. Janikow [7], Olaru and Wehenkel [11] have introduced other algorithms for fuzzy decision tree induction. Suárez and Lutsko [14], and Wang and Chen [18] have presented some optimizations of fuzzy decision trees, while Nikravesh [10] presented evolutionary computation based multi-aggregator fuzzy decision trees. As successors, the Pattern Tree (PT) [5] and fuzzy operator tree [19] algorithms offer several improvements over the previous fuzzy based algorithms. The PT algorithm is able to retain the traditional tree like hierarchical structure. This algorithm provides opportunity to use different aggregation operators and similarity metrics with the aim to obtain better results. This is very important in fuzzy modelling because decisions could be made on basis on multiple-criteria group decision making. There were efforts, mainly in the area of rule fuzzy induction, to build such algorithms, like in the research made by Kóczy, Vámos and Biró [8], where they have proposed fuzzy theory concepts to model complex datasets with different aggregation operators including triangular intersections and average aggregation operators. The aggregation operators provide opportunity for multiple-criteria group decision making. This can be also achieved with the Weighted Pattern Tree (WPT) algorithm [6], which additionally gives weights how well a given pattern tree represent the corresponding class. Beside the use of fuzzy aggregation operators and similarity metrics, the algorithm could be applied by using various MFs.

Therefore, in this paper we experimentally evaluate the influence of the Trapezoidal, Triangular and Gaussian MFs on the WPT algorithm using different number of MFs per attribute over three ecological datasets. These datasets contain 10 input attributes detailing the abundances of the diatoms found in lake and one output attribute that describes the ecological status of the lake using certain parameters (in our case: Conductivity, pH and Saturated Oxygen). Furthermore, we investigate how the complexity and resistance to over-fitting influence on the model's performance. Models with different depths (5 and 10) are used for this purpose, as well as the four variants of the PT algorithm that are described in [5]. These four model variants are obtained by using various settings in the algorithm. We evaluate the resistance to over-fitting of the WPT algorithm by calculating the Root-mean squared error (RMSE) between the two evaluation procedures (in the first experiments the entire data set is used for both training and testing, while in the second experiments a cross validation is made). In

order to ensure that the improvement in the performance is statically significant, we use statistical significance test. For this purpose, we use a two-stage procedure that combines the Aligned Friedman [3] test and post-hoc Hommel test [4], as described in [2].

The rest of the paper is organized as follows: Sect. 2 provides description of the WPT algorithm and its main building blocks: definitions of the similarity metric, aggregations operators, as well as the three MFs that are used. In Sect. 3, the dataset description as well as the experimental setup procedures are described. Section 4 presents the experimental results as well as the conclusions and discussions from this experimental evaluation. Section 5 concludes the paper and outlines our future work.

2 Weighted Pattern Tree Algorithm

The WPT algorithm uses the same concepts as the pattern tree algorithm. The similarity metric is also used to weight the confidence of a particular tree to predict a given class. Before this can be done, the WPT algorithm builds a tree model using aggregation operators and fuzzy terms from the input dataset. The input values for these fuzzy terms are obtained by fuzzification as every algorithm based on fuzzy theory requires, which could be done using different mathematical distributions. In this section, first we will describe the main concepts that are used in the algorithm, and then we will present how the WPT is induced.

In fuzzy theory, MFs are used to transform the crisp values or to transfer from the classical domain to fuzzy domain by a process known as fuzzification. This is an important part of the WPT algorithm, because the fuzzification has influence on the performance of the algorithm. The polygonal MFs (triangular and trapezoidal MFs) have advantages of simplicity, but in many cases datasets consist of smoothed and nonzero values, which are handled much better by Gaussian MF and thus more accurate models are produced. Besides the impact of the MF on the model's performance, also it is important how many MF are used per attribute. Namely, each attribute produce a predefined number of fuzzy terms that could be linguistically labelled. Therefore, in this paper we don't just experimentally evaluate the different types of MFs, but also we make experiments by using different number of MFs per attribute.

Another parameter that plays an important role in increasing the performance of the algorithm is the similarity metric. Therefore, it is important to consider metrics that will reflect the dataset properties. In this paper, we use the similarity metric proposed in [5], which could be used to calculate the similarity between two fuzzy sets A and B defined in the universe of discourse U as a complement of the root mean squared error metric. The similarity is computed as

$$
Sim_{RMSE} = 1 - \sqrt{\frac{\sum_{i=1}^{n} (\mu_A(x_i) - \mu_B(x_i))^2}{n}}, \tag{1}
$$

where $\mu_A(x_i)$ and $\mu_B(x_i)$ are the membership degrees of an element x_i in the fuzzy sets A and B, respectively. The larger the similarity is, the more similar the fuzzy sets are.

The similarity metric could be used to evaluate the degree of similarity between a given fuzzy set and the class attribute. These fuzzy sets could be the initial sets that are obtained with fuzzification. However, they often do not provide best description of a given class. For that purpose, fuzzy aggregation operators are used in order to combine the fuzzy sets. There are several categories of aggregation operators, but mainly triangular norms are used as in many other fuzzy algorithms. In this research paper, we use the Algebraic AND and Algebraic OR aggregation operators. However, in future our focus can shift towards additional aggregation operators, since they also play an important role in producing more accurate models.

The induction of WPT starts by fuzzification of the dataset by using a particular MF. Each attribute is presented by a given number of fuzzy terms. For each of the fuzzy terms, a pattern tree model is generated. These pattern trees are called as primitive trees, and also we will refer to them as trees at level 0. However, they are too simple and could not provide accurate results in making predictions. Thus, by using aggregation operators, these primitive trees could be aggregated, thus obtaining more complex models. For that purpose, first, for each primitive tree the similarity between the fuzzy set that corresponds to the particular tree and the fuzzy set for the class attribute is calculated. The best primitive tree, which is the tree for which highest similarity is obtained, is further aggregated with the remaining primitive trees. In this way, the candidate trees are obtained. We will refer to these trees as trees at level 1. The candidate tree for which highest similarity is calculated is further aggregated with the remaining primitive tress, thus obtaining the candidate trees at level 2, and this is repeated until the tree reaches the predefined depth. In this way, for each class we obtain separate model tree, and for each of these trees a corresponding weight is assigned based on the similarity between the fuzzy set that corresponds to the tree and the fuzzy set for the class attribute. The model trees that are obtained in this way are simple trees. Besides aggregation of the best candidate tree with the primitive trees that are not used for making that candidate tree, also there is other possibility where besides the primitive trees (trees at level 0), also the trees at the remaining levels (levels 1, 2, etc.) could be considered in the aggregation. In this way, we can distinguish simple models, where in the aggregation only the trees at level 0 are aggregated with the best candidate tree, and general models, where the trees at all levels are considered in the aggregation. The difference between the PT and WPT algorithm is that in WPT besides generation of a model tree for each class, also a corresponding weigh is associated to that models. More details about the induction of the PT and WPT can be found in [5] and [6], respectively.

3   Dataset Description and Experimental Setup

The ecological datasets that are used to experimentally evaluate the influence of the shape, the number of MFs per attribute and the four different model variants are obtained from a real measured dataset collected within an EU project for ecological assessment of Prespa Lake [15]. During the time period of 16 months within the project timeframe, several physico-chemical parameters are measured. This dataset also contains measurements regarding the biological aspects of the lake, through measurements

of the diatoms' abundances. These diatoms live in certain conditions, which are defined with the range of the physico-chemical parameters. Using the measured data, it is possible to find the relationship between the diatoms and the physico-chemical parameters in the environment. In this way, we may find out what are the required conditions in the environment in which a given diatom can survive. In this study, from these physico-chemical parameters, we consider three environmental stress factors whose classification systems can be found in the ecological literature [9, 16, 17]. Conductivity [9], pH [9, 17] and Saturated Oxygen [17] classification systems are directly related to some measured diatom species in a similar manner like the examples found in the ecological literature [16]. Since we are using a single target classification algorithm, which could be used for prediction of a single attribute, while from the measured dataset we consider three physico-chemical parameters, therefore we make a separate dataset by considering each of these three parameters as a single target attribute. The diatom ecological datasets consists from 10 input numeric attributes that represent the relative abundances of the 10 most abundant diatoms in the samples, and one target attribute that may obtain from 4 to 6 nominal values (different number of values for each dataset based on the considered target attribute).

As we mentioned previously, in this paper we evaluate these three ecological datasets and the influence of different number of MFs per attribute (3, 4, 5, 10, 20, 30, 50 and 100). The complexity and resistance to over-fitting of the models are evaluated by using the four different variants of WPTs, which could be simple weighted pattern trees (SWPT) and general weighted pattern trees (GWPT) with depth 5 and 10. The four variants of the WPTs are denoted with: SWPT5 (SWPT with depth 5), SWPT10 (SWPT with depth 10), GWPT5 (GWPT with depth 5) and GWPT10 (GWPT with depth 10). These four WPT variants lead to different properties and performances of the obtained models. For example, trees with higher depth (depth 10) would lead to more complex models but may have higher predictive power, while trees with lower depth (depth 5) are less complex but may have lower predictive power.

The descriptive classification accuracy, denoted as "Train" in the experimental results, shows how well the model tree suits to the training data, where the entire data set is used for both training the model and testing its accuracy. However, in this way we will obtain high results for the models that are over-fitted. Therefore, it is needed to estimate how accurate the model is for unseen data. Therefore, we also use the predictive classification accuracy, denoted as "Test" in the experimental results, by making 10-fold cross validation.

In order to confirm the statistical significance of the obtained results, the two-stage procedure proposed in [2] is employed, which combines the Aligned Friedman test [3] and post-hoc Hommel test [4]. In the first stage, non-parametric Aligned Friedman test [3] is used with so called aligned ranks, which is recommended when the number of experiments is not too large. The average rank according Aligned Friedman test is calculated by using Eq. 2, where k represents the number of different settings variants that could be used, while n is the number of datasets used in the experiments.

$$Rank = \frac{(k-1)[\sum_{j=1}^{k} X_j^2 - (kn^2/4)(kn+1)^2]}{\{[kn(kn+1)(2kn+1)]/6\} - (1/k)\sum_{i=1}^{n} X_i^2} \quad (2)$$

The average rank also considers the total rank for each of these parameters (k and n), where X_i is the total rank for the i-th dataset, while X_j is the total rank for j-th settings variant, respectively. For more information regarding the total rank, the reader is referred to [3]. In ER1 experiment, the number of MFs per attribute is fixed, and the average rank is calculated by using the three types of MFs and the four model variants, so $k = 12$. Similarly, in ER2 experiment, the type of MFs is fixed so $k = 32$, while in ER3 experiment the model variant is fixed so $k = 24$. In this way, the average rank for each dataset is calculated, and also the average rank over all dataset is calculated by using Eq. 2.

Next, the obtained rank is compared for significance with a chi-squared distribution for k - 1 degrees of freedom. The p-value is computed using normal approximations [1], and if the null hypothesis is rejected, usually with high level of significance, we can proceed with the post hoc Hommel test [4]. Since the Aligned Friedman test does not examine the difference among the settings variants, it only shows differences among datasets. For this purpose, pairwise comparison is performed with a post-hoc procedure and a control variant is selected, which is the variant with highest rank, as it is indicated in [2]. The calculated p-value is compared with an appropriate level of significance, usually 0.05, to compensate for multiple comparisons. Because the post-hoc procedure adjusts the level of significance for each comparison, adjusted p-values are recommended to be used in order to make fair comparison among p-values.

4 Experimental Results

In this section, the experimental results are presented for the four model variants (SWPT5, SWPT10, GWPT5, GWPT10), as well as three types of MFs by using different number of MFs per attribute. The descriptive and predictive classification accuracies are determined, and also the RMSE between these two measures is calculated in order to estimate the resistance to over-fitting.

4.1 Performance Analysis

In this section, first, we present the experimental results obtained for the Conductivity dataset (see Table 1). The experiments with the triangular MF are characterized with highest peak of descriptive performance when we are using thirty MFs per attribute, while best predictive performance is obtained with ten MFs per attribute. If we investigate the different WPTs variants that we have employed, we may found out that GWPT5 and GWPT10 have highest predictive and descriptive classification accuracy, which is confirmed by the average accuracy.

When we are using the trapezoidal MF, the models with highest descriptive and predictive performance are achieved with hundred MFs per attributes, except for the descriptive analysis for GWPT5 variant when the best descriptive accuracy is achieved with twenty MFs per attribute. But, the average accuracy is the same as in the previous experiment.

For Gaussian MF, the highest value for the model's descriptive accuracy is when thirty MFs per attribute are used, while the predictive performance settle the highest peak between four and five MFs per attribute. In the case of Gaussian MF, the simple and general WPT variants are similar between each other, but it is interesting that the average accuracy, again, is highest for the descriptive performance for the GWPT10

Table 1. Evaluation results for the Conductivity dataset by using different number of MFs per attribute. Train denotes descriptive classification accuracy, while Test denotes predictive classification accuracy. Underlined results show the models with highest descriptive classification accuracy, while bolded results show the models with highest predictive classification accuracy.

Triangular membership function										
	3	4	5	10	20	30	50	100	Avg	RMSE
Train[a]	71.56	72.94	75.23	78.44	77.06	76.61	75.69	72.48	75.00	6.82
Test[a]	69.09	70.04	**71.86**	68.61	66.75	67.25	69.05	69.09	68.97	
Train[b]	73.39	72.94	75.23	79.36	78.90	79.82	76.15	74.31	76.26	8.73
Test[b]	69.09	69.59	**71.41**	68.18	65.41	66.77	70.43	66.82	68.46	
Train[c]	71.10	74.77	75.23	77.98	76,15	76.61	74.31	72.48	74.83	5.93*
Test[c]	69.09	70.95	71.84	**72.36**	66.30	66.32	71.36	69.55	69.72	
Train[d]	72.02	74.77	75.23	78.90	77.98	81.19	76.15	74.77	76.38	8.12
Test[d]	69.07	69.13	70.93	71.90	66.77	65.84	**72.29**	68.18	69.27	

Trapezoidal membership function										
	3	4	5	10	20	30	50	100	Avg	RMSE
Train[a]	74.31	73.85	72.94	77.06	77.52	77.98	76.15	78.90	76.09	8.10
Test[a]	69.98	67.73	65.41	71.86	63.53	68.59	68.59	**72.27**	68.50	
Train[b]	73.85	73.85	72.94	79.82	79.36	81.19	79.82	82.11	77.87	10.31
Test[b]	69.07	66.36	64.48	70.95	64.46	67.68	68.14	**73.18**	68.04	
Train[c]	73.39	74.31	76.61	78.90	80.28	77.98	76.15	78.90	77.06	8.53
Test[c]	69.98	69.09	68.20	71.43	64.03	69.05	69.03	**74.13**	69.37	
Train[d]	73.85	74.77	76.15	80.28	81.19	81.19	80.28	81.19	78.61	11.03
Test[d]	69.98	67.73	67.29	70.06	63.12	67.68	68.59	**72.27**	68.34	

Gaussian membership function										
	3	4	5	10	20	30	50	100	Avg	RMSE
Train[a]	71.56	74.77	74.31	77.52	76.61	79.36	75.69	73.85	75.46	8.03
Test[a]	68.14	70.95	**73.27**	68.64	67.29	65.45	68.64	64.89	68.41	
Train[b]	72.48	74.77	74.31	77.06	77.98	81.19	76.61	76.61	76.38	8.55
Test[b]	69.09	**71.41**	71.39	67.73	64.91	68.18	70.45	67.21	68.80	
Train[c]	73.39	74.77	75.23	77.52	75.69	79.36	75.69	73.85	75.69	7.38
Test[c]	69.55	69.52	**72.79**	68.64	69.13	66.36	70.00	65.80	68.97	
Train[d]	73.39	74.77	75.23	77.98	78.44	81.19	76.61	76.61	76.78	8.77
Test[d]	69.07	70.00	**71.86**	67.25	66.32	69.55	70.45	65.37	68.73	

[a]SWPT5
[b]SWPT10
[c]GWPT5
[d]GWPT10

variant, while the predictive performance is best with GWPT5. If we examine the resistance to over-fitting, we can note that best value is achieve with GWPT5 variant using triangular MF with value of almost 6% error. In the other experiments, various models obtain 2 to 4% higher error when it comes to resistance to over-fitting.

In Table 2, the experimental results obtained for the pH dataset are presented. The first part of the table depicts the influence of the triangular MF for which best descriptive results are achieved using high number of MFs per attribute (fifty and hundred). On the other side, for predictive accuracy, the models with low number of MFs per attribute (between three and five) obtain best results.

Table 2. Evaluation results for the pH dataset by using different number of MFs per attribute. Train denotes descriptive classification accuracy, while Test denotes predictive classification accuracy. Underlined results show the models with highest descriptive classification accuracy, while bolded results show the models with highest predictive classification accuracy.

Triangular membership function

	3	4	5	10	20	30	50	100	Avg	RMSE
Train[a]	58.26	61.01	62.84	62.84	62.39	62.84	65.60	65.60	62.67	14.78
Test[a]	56.30	**57.12**	56.73	48.92	41.49	45.71	48.42	43.85	49.82	
Train[b]	59.63	61.47	62.84	64.68	66.51	69.27	70.64	71.56	65.83	17.58
Test[b]	56.30	57.14	**57.16**	50.74	43.31	43.90	49.81	47.06	50.68	
Train[c]	60.09	61.93	61.01	63.30	62.39	62.84	66.97	65.60	63.02	13.65
Test[c]	**55.82**	53.90	54.44	48.44	44.22	50.80	48.87	47.10	50.45	
Train[d]	61.47	63.30	61.01	65.14	66.51	69.27	73.85	72.48	66.63	17.00
Test[d]	**55.82**	54.39	55.80	50.74	45.13	49.42	50.69	48.44	**51.30**	

Trapezoidal membership function

	3	4	5	10	20	30	50	100	Avg	RMSE
Train[a]	59.17	60.55	61.93	62.39	65.60	67.43	66.06	65.14	63.53	14.26
Test[a]	53.01	**54.85**	54.46	46.58	51.13	50.24	48.42	44.29	50.37	
Train[b]	59.17	60.55	61.93	63.76	66.97	68.35	70.64	72.94	65.54	16.83
Test[b]	53.01	54.85	**56.73**	47.97	50.17	49.35	47.03	46.15	**50.66**	
Train[c]	60.55	60.09	63.30	62.39	65.60	65.60	66.06	65.14	63.59	14.82
Test[c]	52.55	55.30	**55.41**	46.10	48.42	47.03	50.30	43.79	49.86	
Train[d]	60.55	60.09	63.76	63.76	68.35	69.27	71.10	73.39	66.28	18.26
Test[d]	52.55	53.90	**58.14**	47.03	48.38	46.15	47.06	46.13	49.92	

Gaussian membership function

	3	4	5	10	20	30	50	100	Avg	RMSE
Train[a]	58.72	57.80	61.93	63.30	62.39	59.63	66.97	65.60	62.04	13.16
Test[a]	**55.35**	53.01	54.89	53.92	47.94	42.51	49.81	45.17	50.32	
Train[b]	58.26	57.34	62.39	63.30	67.43	63.76	72.02	71.56	64.51	16.14
Test[b]	55.35	**55.80**	55.37	52.99	48.38	41.10	51.65	46.56	**50.90**	
Train[c]	59.17	57.80	55.50	63.30	61.01	59.63	66.06	64.68	60.89	12.22*
Test[c]	53.05	52.51	50.28	**53.46**	48.40	45.26	47.49	47.49	49.74	
Train[d]	59.63	57.80	56.42	63.76	66.06	65.60	72.02	71.56	64.11	15.82
Test[d]	53.05	**56.71**	52.10	52.53	48.38	43.44	50.24	48.40	50.61	

[a]SWPT5
[b]SWPT10
[c]GWPT5
[d]GWPT10

The simple WPTs obtain best predictive accuracy, while general WPTs are better in descriptive analysis. The average accuracy in the overall performance analysis for the triangular MF, placed the GWPT10 and GWPT5 among the best for building diatom ecological models. WPT with depth 10 are better in ecological modelling than the WPT with depth 5 based on the model's descriptive accuracy.

The next MF is the trapezoidal function, for which higher number of MFs per attribute (from thirty till hundred) give better descriptive performance, while the models that have low number of MFs per attribute have best predictive performance. Regarding the WPT variant, the models with depth 5 are worse in both descriptive and predictive analysis, while both simple and general models with depth 10 are better in the experimental evaluation. This is confirmed by the average performance analysis, which puts the SWPT10 as the best model with highest predictive accuracy, while GWPT10 is best in the descriptive analysis.

And finally, the Gaussian MF doesn't change the pattern that we found for this dataset. It attains high descriptive accuracy using high number of MFs per attributes (exactly fifty) and low number of MFs per attribute for predictive accuracy (between three and ten). When it comes to depth analysis, again, simple and general WPT with depth 10 are best for obtaining models with high descriptive and predictive accuracy. This is confirmed with the obtained average classification accuracy, but compared to the triangular MF, SWPT10 are best in the case of Gaussian MF. The analysis of the RMSE regarding the resistance to over-fitting over this dataset, shows lower resistance to over-fitting compared to the Conductivity dataset (twice much higher error of 12.22%), and the best model variant to build such models is using GWPT5. The other variants are having 3 to 5% lower resistance to over-fitting.

The experimental results for the last dataset that we used, the Saturated Oxygen dataset, are presented in Table 3. Here, the triangular MF continues the trend of separation between the models regarding their descriptive and predictive accuracy, when it comes to the number of MFs per attribute. In this case, the descriptive accuracy is higher with higher number of MFs per attribute (more than twenty), while the models obtained by using low number of MFs per attribute (between three and five) are having highest value for predictive accuracy. There is no big difference when using simple or general WPT, even with different depths. That's why there is not much difference in the average accuracy, where SWPT5 slightly overruns the other variants for predictive analysis, while GWPT10 is best in the descriptive analysis.

If we examine the trapezoidal MF, the trend of separating the train and test performance with different number of MFs per attribute stops here. Here, the best models are obtained when the number of MFs per attribute is higher than 5. When it comes to model variants, the results are very close, and there is not much difference between the models with depth 5 and 10. The average accuracy shows that the best model with highest descriptive accuracy is the same as for the triangular MF, while SWPT10 is best for obtaining models with high predictive performance.

The best model according to the descriptive performance is the same when using the Gaussian MF, while best average predictive performance is obtained by using the GWPT10 variant. Here, there is not much difference between the descriptive and predictive performance when it comes to number of MFs per attribute. It is noticeable that for the modes obtained by using more than 5 MFs per attribute, better descriptive

as well as predictive accuracy is achieved. Here, again, there is no noticeable difference regarding the WPT variants, both simple and general models with different depths can be used and they obtain very similar results.

Table 3. Evaluation results for the Saturated Oxygen dataset by using different number of MFs per attribute. Train denotes descriptive classification accuracy, while Test denotes predictive classification accuracy. Underlined results show the models with highest descriptive classification accuracy, while bolded results show the models with highest predictive classification accuracy.

Triangular membership function										
	3	4	5	10	20	30	50	100	Avg	RMSE
Train[a]	60.70	61.19	63.68	65.67	65.17	62.19	<u>67.66</u>	63.18	63.68	9.53*
Test[a]	**57.00**	56.50	55.50	53.50	52.50	53.50	56.00	53.00	**54.69**	
Train[b]	60.70	61.69	64.68	67.16	66.17	66.17	68.66	<u>69.15</u>	65.55	11.99
Test[b]	55.50	**56.50**	**56.50**	52.50	54.50	52.00	54.50	52.50	54.31	
Train[c]	60.70	59.70	60.20	64.68	<u>68.66</u>	62.69	67.66	63.18	63.43	11.76
Test[c]	**57.00**	54.50	53.00	51.00	52.00	48.00	52.50	53.00	52.63	
Train[d]	60.20	59.70	61.69	66.67	69.65	<u>69.65</u>	68.66	68.66	<u>65.61</u>	13.70
Test[d]	**56.00**	54.50	55.00	53.00	53.50	49.00	52.50	52.00	53.19	

Trapezoidal membership function										
	3	4	5	10	20	30	50	100	Avg	RMSE
Train[a]	60.20	60.70	62.19	65.17	66.17	61.69	66.67	<u>67.66</u>	63.81	11.04
Test[a]	53.00	53.00	55.00	55.00	**56.00**	46.00	**56.00**	52.00	53.25	
Train[b]	60.20	61.19	63.68	67.66	<u>70.65</u>	64.68	69.15	69.15	65.80	13.07
Test[b]	53.00	52.00	55.50	57.00	**58.00**	46.50	54.00	51.00	**53.38**	
Train[c]	60.20	61.19	62.19	66.17	66.17	61.69	67.16	<u>67.66</u>	64.05	12.01
Test[c]	53.00	53.00	**56.00**	54.50	55.00	43.50	54.50	52.00	52.69	
Train[d]	60.70	61.69	63.68	67.66	68.66	64.68	69.65	<u>69.15</u>	<u>65.73</u>	13.49
Test[d]	53.50	52.00	56.00	**56.50**	**56.50**	44.50	53.50	51.50	53.00	

Gaussian membership function										
	3	4	5	10	20	30	50	100	Avg	RMSE
Train[a]	56.22	57.71	57.71	63.18	<u>65.67</u>	<u>65.67</u>	<u>65.67</u>	<u>65.67</u>	62.19	10.95
Test[a]	52.00	52.00	49.00	56.00	50.00	**53.00**	51.00	**53.00**	52.00	
Train[b]	56.72	58.21	58.71	65.17	67.66	<u>70.15</u>	69.15	69.65	64.43	13.50
Test[b]	53.00	52.50	49.50	**54.00**	52.50	53.00	51.00	51.00	52.06	
Train[c]	58.71	60.20	61.19	61.69	<u>68.16</u>	65.67	65.17	65.67	63.31	15.19
Test[c]	52.00	52.50	**53.00**	51.50	50.50	41.50	42.50	**53.00**	49.56	
Train[d]	58.71	60.20	58.71	60.70	69.65	<u>70.15</u>	69.15	69.65	64.61	16.52
Test[d]	54.00	53.50	52.50	50.00	51.50	44.00	43.00	**56.00**	**54.31**	

[a]SWPT5
[b]SWPT10
[c]GWPT5
[d]GWPT10

For the Saturated Oxygen dataset, the resistance to over-fitting is better than for the pH dataset, and it is worse compared to the Conductivity dataset, peaking lowest error of 9.53%.

The in-depth analysis found mixed patterns which cannot be distinguished. In order to test/report statistical significance of the results, next, we present the results from the two-stage procedure.

4.2 Ranking the MFs and WPT Variants

In this section, we perform ranking by using different number of MFs per attribute, different types of MFs, and different WPTs variants. The ranking is done by using the two stage procedure presented in [2] in order to measure the statistical significance of the obtained results. This procedure combines the Aligned Friedman test [3] and the post-hoc Hommel test [4]. The results from the ranking are presented in Table 4.

Table 4. The average ranks over the three datasets by using different settings (different number of MFs per attribute, different types of MFs and WPTs variants). The best rank is bolded and the model with best average rank is taken as control model for the post-hoc Hommel test. The rejected null hypothesis based on the Hommel adjusted p–values are underlined, and they correspond to statistically significant differences between the examined and the control model.

Different settings	Conductivity		pH		Saturated Oxygen		All datasets	
ER1	Train	Test	Train	Test	Train	Test	Train	Test
3	86.45	40.75	82.16	22.66	83.62	**36.21**	290.5	91.6
4	73.00	41.41	78.75	18.41	78.96	42.58	230.4	93.2
5	65.66	**33.92**	69.12	**16.92**	69.25	37.75	204.6	**84.4**
10	24.04	36.83	52.62	51.92	46.96	40.00	126.7	132.8
20	28.71	82.75	38.62	68.67	22.96	43.75	91.6	194.7
30	**14.16**	68.25	37.37	74.08	39.67	78.95	97.5	221.8
50	46.54	35.08	**13.38**	55.96	**21.79**	53.14	**72.6**	152.7
100	49.42	49.00	15.96	78.37	24.79	55.62	82.8	184.6
ER2	Train	Test	Train	Test	Train	Test	Train	Test
Triangular	62.95	**43.53**	38.33	**44.17**	44.98	**35.75**	144.6	**121.4**
Trapezoidal	**28.62**	53.00	**37.87**	52.98	**38.98**	43.58	**103.8**	148.3
Gaussian	53.92	48.97	69.30	48.34	61.53	66.17	185.0	163.7
ER3	Train	Test	Train	Test	Train	Test	Train	Test
SWPT5	71.18	50.71	66.77	52.87	69.25	40.20	204.7	142.6
SWPT10	35.98	58.25	32.12	**39.12**	31.54	**37.75**	98.12	**131.1**
GWPT5	61.54	**33.10**	68.54	57.27	61.92	64.71	192.5	157.5
GWPT10	**25.29**	51.94	**26.56**	44.73	**31.29**	51.33	**88.6**	146.7

The results from this ranking confirmed what we have concluded in the previous section: higher number of MFs per attribute (thirty for Conductivity and fifty for pH and Saturated Oxygen datasets) are best for obtaining models with higher descriptive accuracy, and these results showed as statistically significant compared to the models built with lower number (three, four and five) of MFs per attribute. Completely different picture can be seen for the models' predictive accuracy, where lower number of MFs per attribute, especially five MFs per attribute, are statistically significant compared to the models that are built with higher number of MFs per attribute (higher than ten). The conducted experiments that include all the datasets confirms this. Another set of experiments that were conducted evaluated different types of MFs. In most cases, the results for the triangular and trapezoidal MFs are statically significant compared to the results for the Gaussian MF. And the final analysis was made to evaluate the different WPT variants, which confirmed some of the discussion points that we made in the previous section. In most of the cases simple and general WPT models with depth 10 achieve best predictive and descriptive performance, and they are statistically significant compared to the models with depth 5. GWPT10 obtains statically significant descriptive performance compared to the models with depth 5, while SWPT10 is the best model in predictive analysis but without statistical significance.

5 Conclusion

In this paper work, we performed extensive experimental evaluation of the influence of the different MFs on the classification accuracy on diatom datasets using the WPT algorithm. Beside the influence of the different MFs on the descriptive and predictive performances of the models, we also investigated the influence of the number of MFs use per attribute and the different WPTs variants that have different complexities on the accuracy of the models. The results from the evaluation revealed some interesting patterns, like best ranked model with highest descriptive classification accuracy can be obtained using high number of MFs per attribute (fifty) in combination with triangular MF and GWPT models with depth 10. On the other hand, models built with lower number of MFs per attribute in combination with Triangular MF and SWPT with depth 5, have highest predictive classification accuracy. These conclusions are based on the outputs obtained from the two-stage procedure for testing the statistical significance of the results.

As future work, we plan to investigate other types of MFs, as well as different similarity metrics and aggregation operators that influence on the classification accuracy of the models.

Acknowledgment. This work was partially financed by the Faculty of Computer Science and Engineering at the Ss. Cyril and Methodius University in Skopje.

References

1. Abramowitz, M.: Handbook of Mathematical Functions. Graphs, and Mathematical Tables, With Formulas, Dover Publications (1974)
2. García, S., Fernández, A., Luengo, J., Herrera, F.: Advanced nonparametric tests for multiple comparisons in the design of experiments in computational intelligence and data mining: experimental analysis of power. Inf. Sci. **180**, 2044–2064 (2010)
3. Hodges, J., Lehmann, E.: Ranks methods for combination of independent experiments in analysis of variance. Annal. Math. Stat. **33**, 482–497 (1962)
4. Hommel, G.: A stagewise rejective multiple test procedure based on a modified Bonferroni test. Biometrika **75**, 383–386 (1988)
5. Huang, Z., Gedeon, T.D.: Pattern trees. In: 2006 IEEE International Conference on Fuzzy Systems, pp. 1784–1791. IEEE (2006)
6. Huang, Z., Nikravesh, M., Azvine, B., Gedeon, T.D.: Weighted pattern trees: a case study with customer satisfaction dataset. In: Melin, P., Castillo, O., Aguilar, L.T., Kacprzyk, J., Pedrycz, W. (eds.) IFSA 2007. LNCS, vol. 4529, pp. 395–406. Springer, Heidelberg (2007). https://doi.org/10.1007/978-3-540-72950-1_39
7. Janikow, C.Z.: Fuzzy decision trees: issues and methods. IEEE Trans. Syst. Man Cybern. **28**(1), 1–14 (1998)
8. Kóczy, L.T., Vámos, T., Biró, G.: Fuzzy signatures. In: EUROFUSE-SIC, pp. 210–217 (1999)
9. Krammer, K., Lange-Bertalot, H.: Die Ssswasserflora von Mitteleuropa 2: Bacillario-phyceae. 1 Teil, pp. 876. Gustav Fischer-Verlag, Stuttgart (1986)
10. Nikravesh, M., Bensafi, S.: Soft computing for perception-based decision processing and analysis: web-based BISC-DSS. In: Nikravesh, M., Zadeh, L.A., Kacprzyk, J. (eds.) Soft Computing for Information Processing and Analysis. Studies in Fuzziness and Soft Computing, vol. 164, pp. 93–188. Springer, Heidelberg (2005). https://doi.org/10.1007/3-540-32365-1_4
11. Olaru, C., Wehenkel, L.: A complete fuzzy decision tree technique. Fuzzy Sets Syst. **138**, 221–254 (2003)
12. Quinlan, J.R.: Decision trees and decision making. IEEE Trans. Syst. Man Cybern. **20**(2), 339–346 (1990)
13. Stroemer, E.F., Smol, J.P.: The Diatoms: Applications for the Environmental and Earth Sciences. Cambridge University Press, Cambridge (2004)
14. Suárez, A., Lutsko, J.F.: Globally optimal fuzzy decision trees for classification and regression. IEEE Trans. Pattern Anal. Mach. Intell. **21**(12), 1297–1311 (1999)
15. TRABOREMA Project: WP3, EC FP6-INCO project no. INCO-CT-2004-509177 (2005–2007)
16. Van Dam, H., Martens, A., Sinkeldam, J.: A coded checklist and ecological indicator values of freshwater diatoms from the Netherlands. Netherlands J. Aquatic Ecol. **28**(1), 117–133 (1994)
17. Van Der Werff, A., Huls, H.: Diatomeanflora van Nederland. Abcoude - De Hoef (1957, 1974)
18. Wang, X., Chen, B., Olan, G., Ye, F.: On the optimization of fuzzy decision trees. Fuzzy Sets Syst. **112**, 117–125 (2000)
19. Yi, Y., Fober, T., Hüllermeier, E.: Fuzzy operator trees for modeling rating functions. Int. J. Comput. Intell. Appl. **8**(04), 413–428 (2009)
20. Yuan, Y., Shaw, M.J.: Induction of fuzzy decision trees. Fuzzy Sets Syst. **69**(2), 125–139 (1995)

A Study of Different Models
for Subreddit Recommendation Based
on User-Community Interaction

Andrej Janchevski$^{(\boxtimes)}$ ⓘ and Sonja Gievska$^{(\boxtimes)}$ ⓘ

Faculty of Computer Science and Engineering, Ss. Cyril and Methodius University,
ul.Rudzer Boshkovikj 16, P.O. 393, 1000 Skopje, Republic of North Macedonia
andrej.jancevski@students.finki.ukim.mk, sonja.gievska@finki.ukim.mk

Abstract. Reddit is a community-oriented social network, where users can pose questions, share their own views and experiences within subreddit communities they have subscribed to, with the possibility that other users might view, rate and comment on their posts. A recommender system plays a crucial role in advancing and steering interactions on social media platforms, and in the case of Reddit, it performs across many levels. This study investigates the potential benefits of social media analytics for improving the quality of recommendations. Five models are proposed and validated, with a particular focus on improving the recommendations of subreddits that might be of interest to a particular user. The results reinforce the notion that capturing and fusing diverse set of features is crucial for confronting the challenges of predicting elusive phenomenon such as user's preferences and interests.

Keywords: Social network analytics · Recommender systems · Data fusion · Network embeddings · Content similarity

1 Introduction

Reddit is a community-oriented social media platform, where users can post links and their own content, rate and comment on other users' postings within thematic communities i.e., subreddits they belong to. The quality and type of submitted user's post adhere to the rules set and are controlled by each subreddit community.

Filtering and relevance-ordering of the posts, comments and subreddits a user is advised to view or join helps both, active and passive participants deal with the overwhelming number of shared content. In the era, when popular social networks silently run intelligent software that acquires huge amounts of private user data infringing on the boundaries of ethics, Reddit prides itself for requesting no personal information from its users as they remain anonymous. In this research, we argue that social media analytics of user-community interaction network and natural language processing of user content may be employed as a basis for a recommender system that still respects user's anonymity.

© Springer Nature Switzerland AG 2019
S. Gievska and G. Madjarov (Eds.): ICT Innovations 2019, CCIS 1110, pp. 96–108, 2019.
https://doi.org/10.1007/978-3-030-33110-8_9

A recommender system plays a crucial role in advancing and steering interactions on social media platforms, and in the case of Reddit it could perform across many levels:

- Subreddit recommendation - new users receive general information about popular subreddits, while veteran users are notified of new communities that are expected to be aligned with user's preferences and interests.
- Post recommendation - personalized ordering of the posts should help each subreddit's subscriber focus on the most relevant posts.
- Comment recommendation - a relevance-ordering of comment threads or sub-threads in line with user's current and past interest.

The scope of this study is limited to the problem of subreddit recommendation. Overall, this research has provided insights into the roles different types of information play in the recommender system under investigation. The analysis has suggested several directions that can be pursued to extend the current research.

The next section follows the most related research to trace the new trends in the field. The five models for improving subreddit recommendations proposed in this paper are discussed in Sect. 3. Section 4 presents the evaluation results of testing the models and points to future research directions that could strengthen our work. Section 5 concludes the paper.

2   Related Work

The research objectives for proposing a new recommender system are drawn from the shortcomings of what is currently being offered by the social media platform in question and the ongoing research in the field of recommender systems. This research follows the line of work of the research groups that have contributed to the discourse on recommender systems.

Jamonnak et al. [7] use the Apriori association rule mining algorithm for discovering subreddits that might be of potential interest to a user, which was incorporated in a web-based application called Recommenddit that allows a user to query and retrieve a set of subreddits closely related to a particular subreddit community. The associations between two subreddits were established based on the overlap of active users participating in both communities. In contrast, our approach for improving the subreddit recommendations uses a large number of diverse set of features, extracted by network analysis of graphs created on the basis of user-subreddit interactions, and text processing of users' postings.

A Reddit recommendation engine, named RedTweet, proposed by Nguyen et al. [11], relies upon text analysis of user's postings on another social network, Tweeter to build user profiles that might shed light on users' interest and preferences. A number of natural language processing techniques and resources have been utilized, from WordNet-based similarity measures to genre classification based upon an augmented genre-tagged Brown Corpus. The top K genres detected in user's profile are used to recommend subreddit articles that are

inline with their interests. A comparative analysis of the performance of their genre-based classification approach against topic classification based on traditional machine learning algorithms points to comparable results. While content similarity of users' posts has been included in our models, the text analysis we perform is limited to Reddit users' postings, without the need for inclusion of external sources of information for identification of user's interests in certain topics.

Another text processing inclined research on community recommendations on Reddit is published recently by Tuomchomtam and Soonthornphisaj [15]. The study uses a number of textual features for clustering subreddit communities by using the Density-based clustering algorithm for applications with noise (DBSCAN) algorithm, followed by a logistic regression algorithm for ranking the subreddits mostly related to a given post. High accuracy of 90% has been achieved during the evaluation of this framework, which is comparable to the performance reported in this paper.

We would like to acknowledge two student projects that incited interesting ideas that were considered in our framework for subreddit recommendations. The first one, explores the use of a modularity based community detection algorithm for clustering subreddits into global communities against which user's subreddit subscriptions are compared [14]; when a match is found, the recommendations for a new subreddit to a user comes from the matching group of subreddit communities. The latter postulated that chronologically-ordered sequence of user's subreddit subscriptions captures the variations, but also the stability of user's interests through time [9] - past user's subscriptions were used for training various deep learning architectures to "learn" to provide recommendations that are inline with past user's subreddit discoveries.

A plethora of other studies on Reddit recommender systems has been reported, providing evidence of a number of alternative approaches for solving the problem Lakkaraju et al. [8], Poon et al. [13], Das et al. [3], Huang et al. [6]. Although, the focus of some of these studies is on recommending posts instead of subreddit communities, they provide insights into the suitability and effectiveness of particular methods and point to a particular type of social network analysis and behavior analytics that could be fruitfully applied in similar contexts and objectives.

While we draw upon some of the ideas presented across studies with objectives related to ours, the models and analysis presented in this paper brings new insights to the discussion and offers promising direction for future research. Comparative performance analysis of five subreddit recommender models proposed in this paper with the related research is challenged by the differences in methodologies, deep learning vs. traditional feature engineering, as well as the size of the datasets, giving a unique advantage to each model in certain situations.

3 Methodology

The following section describes the details of the models of Subreddit recommender system we have experimented with. Three models that rely upon information extracted from diverse sources were evaluated as a basis for the proposal of two more robust fusion-ensemble style models. This study highlights the relevance of both, network analysis of graphs created on the basis of user-subreddit interactions, and text processing of users' postings for recommending subreddits, which prove to highly coincide with user interests.

The system[1] was implemented using Python 3.6, with the ScikitLearn [12], NLTK [1], and NetworkX [5] packages.

3.1 Dataset

The Reddit dataset collected for previous research by Lakkaraju et al. [8] and made publicly available as part of the Stanford SNAP dataset collection[2] was used. The dataset was collected with a specific objective in mind, namely, analysis of the popularity of users' postings containing images, which proved to be easier to follow in resubmissions by using the reverse image searching tool designed for Reddit.

Each image in the dataset is associated with its title and URL, as well as a number of metadata related to the post: the date and time when the post was created, the username of the person who submitted the image, the subreddit to which it had been posted and the number of upvotes and downvotes the submission had received. In addition, the entire comment threads attached to every post had been retrieved, with each comment complemented with similar metadata as the posts. In total, the dataset contains 132,307 image posts, with 16,736 unique images. Additionally, the dataset includes almost 5 million comments and 250 million upvotes/downvotes. Submissions from 63 thousands users to 867 different subreddits are present in the dataset.

The original dataset of the post submissions can be downloaded in a convenient text format, however the comment threads attached to each post are not parsed and are given in their original HTML format. To this end, parsing and preprocessing was required to extract the relevant information from user comments. For this task, the Beautiful-Soup HTML parser was utilized and the following information were extracted for each comment: username of the comment author, name of the subreddit to which the original post had been published, timestamp of the posting and the textual content of the comment. Similarly, for each post item in the text collection, the following fields were deemed relevant for our research objective and were extracted: username of the post author, name of the subreddit where it had been posted, the post title, and the timestamp of the posting.

---

[1] The complete source code is available at: https://github.com/Bani57/subredditRecommender.

[2] http://snap.stanford.edu/data/web-Reddit.html.

3.2   User-Subreddit Interaction Networks

At the outset of our exploration, we wanted to map the information between subreddits and users in both directions. Namely, to each subreddit, a list of every user who had made a post or left a comment is attached, conversely to each user, a list of all subreddits the user has shown involvement in is associated. In a total of 847 different online subreddit communities, 658,254 unique users have been active, with a total of 1,266,122 users' postings. While the number of communities is not too large and is computationally viable to process, the number of users and interactions was decided to be reduced.

Close inspection of user activities within subreddits reveals the fact that the majority of users' interactions happen within a very small group of subreddits, while the rest of the activities could be described as activities of leaving a single comment in a single subreddit. It is neither necessary nor valid to keep these interactions as no recommendation can be made for a user with so little data. Data reduction was performed in the following fashion: (1) users showing activity in at least 10 different communities were preserved and (2) two users are set to be linked only if they have 7 subreddits in common. These values were derived from mean and median user-subreddit interaction statistics. The process of data reduction was performed in parallel with the creation of two interaction networks, as a representation of the links between users and subreddits.

Short description of the two interaction network follows:

– **User-subreddit-User (UsU)** network - The nodes of the UsU network represent users, while an edges links two users only if they have at least 7 subreddits in common. The number of overlapping subreddits is associated as an edge weight, while the number of subreddits a user is active is represented as a node weight;

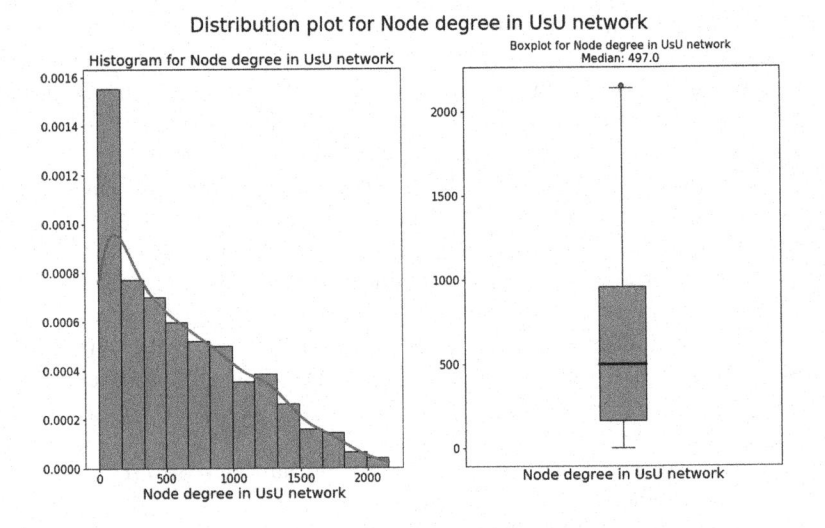

Fig. 1. Histogram and boxplot for node degrees in the UsU network.

- **Subreddit-user-Subreddit (SuS)** network - The nodes of the SuS network represent subreddits, while an edge link two subreddits only if they have at least 1 user in common i.e. a user is subscribed and active in both subreddit communities. The weight of an edge represent the number of users they have in common, while the number of active users in a community is associated as a node weight;

Encoding the activity of users within subreddit communities in a graph format was a prerequisite for the subsequent process of network analysis, which started with investigating the basic properties of the two created networks.

The UsU network is more voluminous than the other, containing 2751 nodes and 845,128 edges, with an average edge density of 22.34%. The SuS network is substantially more compact, containing 847 nodes and 22,940 edges with an average edge density of 6.4%. Degree distribution plots of nodes belonging to the UsU and SuS are shown in Figs. 1 and 2, respectively. Figures 3 and 4 visualize samples from UsU and SuS interaction networks, respectively. The power-law-like degree distributions are not too surprising, as Reddit is a typical scale free network in which only few users and communities achieve global popularity.

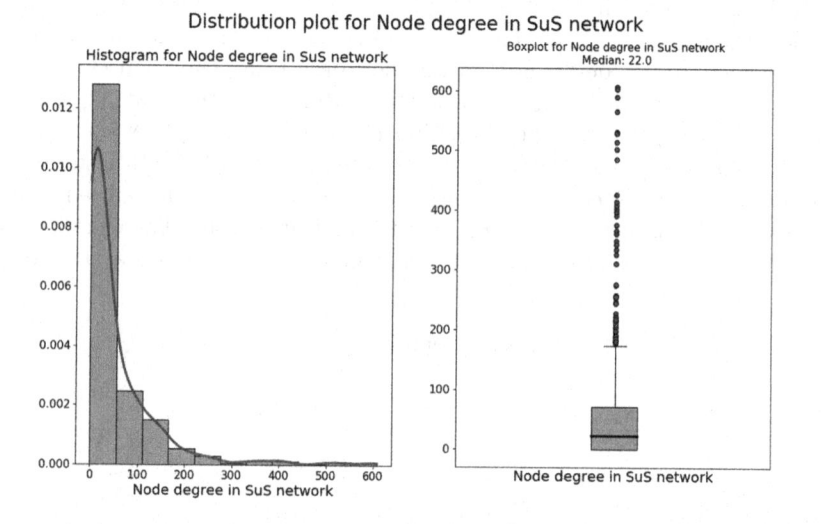

Fig. 2. Histogram and boxplot for node degrees in the SuS network.

3.3 Model Design

Five models for subreddit recommendation are discussed in this paper, each one relies on a collection of diverse type of features that are hypothesized to be important for discriminating between "good" and "bad" subreddit recommendations.

Network Statistics Model. The underlying assumption for the first model is that the association between a user and a subreddit could be based upon the similarities between the network properties of subreddits a user belongs to and the similarities between the "roles" a user has in those communities. It was expected that this particular knowledge is encoded in the two user-subreddit interaction networks we have initially created.

The set of features in this model consist of several standard network statistics, calculated for each node in the UsU and SuS networks, which includes: node degree (weighted and unweighted), degree centrality, closeness centrality (weighted and unweighted), betweenness centrality (weighted and unweighted), clustering coefficient (weighted and unweighted), HITS hub score and PageRank score (weighted and unweighted).

A prediction of a strong connection between a user and a subreddit implies close connections in the SuS network between this particular subreddit and the other subreddit the user belongs too. Conversely, if a user shows a strong sense of belonging to a certain community, then he and the users currently active in this group should exhibit strong ties in the UsU network. To capture these underlying assumptions, average distances, expressed as hop counts and weighted distances, between the corresponding nodes in the two interaction networks were included as additional features in this model.

Following the idea presented in Sundaresan et al. [14], modularity-based community detection was performed on our networks using the Clauset-Newman-Moore greedy modularity maximization algorithm Clauset et al. [2] available in the NetworkX package. The algorithm has detected 1965 user communities in the UsU network, and 292 subreddit communities in the SuS network. The identification of the detected user communities in UsU, and the subreddit communities in the SuS network were added to the features set. The existence of a large number of communities indicates that many of the nodes may be isolated

User-subreddit-user network

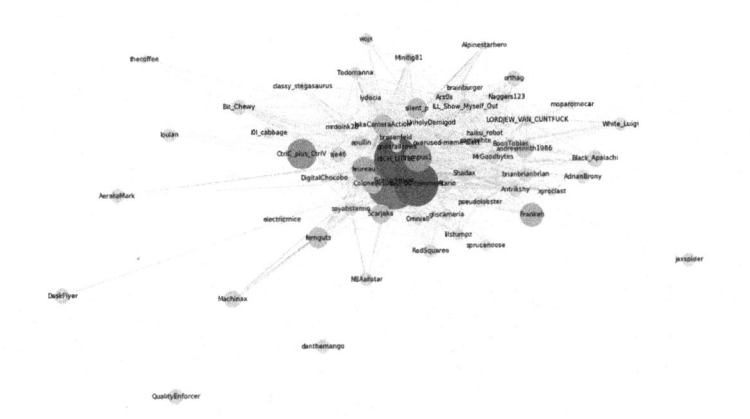

Fig. 3. Visualization of a sample of the UsU network.

or linked to at most one other node, which is not surprising considering the scale-free type of network model the two networks belong to (Figs. 3 and 4). The final set of features for this model included 30 network-specific features in total.

Node Embeddings Model. Processing large networks is computationally-expensive, and in the last decade, more efficient approaches to network analysis based on network embeddings are proposed as a suitable vector representation of graph properties.

Node2Vec algorithm, proposed by Grover and Leskovec [4] was utilized for both, SuS and UsU networks, to generate node embeddings for each user and subreddit. Context window size was set to 5, with 30 random walks of length 50 used to generate node embedding vectors of dimension 500. By simply concatenating the user and subreddit embeddings, a feature vector of length 1000 was constructed for this model.

Subreddit-user-Subreddit network

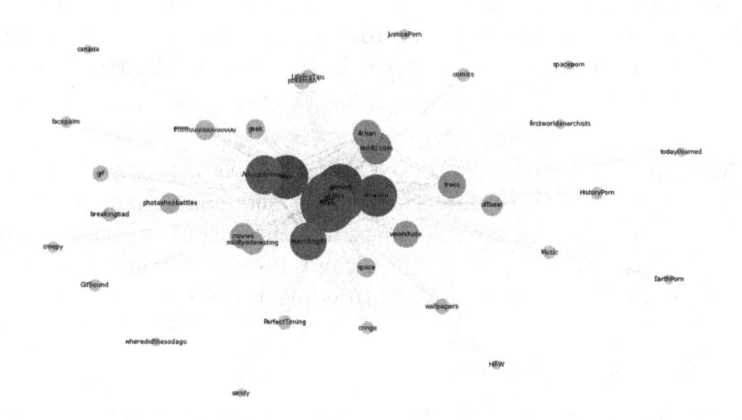

Fig. 4. Visualization of a sample of the SuS network.

Content Similarity Model. One might argue that sometimes a user might be tempted to join an existing subreddit community because of their personal interest in the content that is posted, even though it may appear that the topic is not related to the topics of other subreddits the user is active on. Under the premise that linguistic analysis of user's posts might shed light on their personal interests, we propose analyzing similarity between user's text postings and the entire textual content posted within a certain Reddit community. It was expected that natural language processing might reveal patterns and relations that are missed by network-related features that previous two models rely upon.

Keywords were extracted from the title and the content of user's postings. Two corpuses were generated, one containing the entire content posted by a user, and the other corpus containing the entire textual content posted within a certain subreddit. Each corpus was tokenized, and TF-IDF metric was calculated for each token i.e., unigram. Tokens with TF-IDF score lower than the mean and median were discarded. The top 30 TF-IDF-ranked tokens or all if the number is less than 30, were selected as keywords.

By doing this, words that are commonly used by a user or a community, but not frequently used by others, were identified. The following examples demonstrate that the selected keywords appear to be the appropriate representation of the content posted in a given community. For example, the keywords selected for subreddit r/atheism include: *atheism, bible, belief, jesus, science, church* etc. while the set differs for subreddit r/GifSound: *sync, gif, sound, sidebar, submission*, etc. On the other hand, the keywords representing the text posted by the Reddit user u/markycapone are: *calorie, muscle, exercize, protein, sport*, etc., which point to some of his interests. To compress the information contained in the keyword lists into a numerical feature vector, the Word2Vec model pretrained by Google [10] was used to calculate the mean word embedding vector of size 300 for each keyword list. The total of 600 embedding values for each user-subreddit pair were used as features in the third model based on textual content similarity.

Fusion Models. Two new multi-source fusion models were designed as a combination of the previous three models. The first one falls under the category of decision-fusion, namely, a linear combination of the outputs of the three models were considered as a majority vote. The second, fuses all features from the three models and produces the output in a feature-fusion like manner.

3.4   Training and Evaluation

A feature vector was constructed for every user-subreddit sample present in the two interaction networks per the specific requirements of each of the three models. One must unambiguously define the classification task at hand: for a given (user, subreddit) tuple, the trained model should predict whether the subreddit is considered to be a "good" or a "bad" recommendation for the user. The feature vector representation the models rely upon were expected to capture the information necessary to distinguish between good and bad recommendations.

In the absence of labeled data, we augmented the dataset with samples representing the negative class i.e., what is considered to be a bad recommendation, by random sampling from the pool of subreddits a user is not subscribed to. The sampling is performed to balance the number of negative samples (bad recommendations) with the number of samples of the positive class (good recommendations), denoted by the links between the communities a user is subscribed to.

The models were trained following the same procedure. The list of subreddits was sorted according to their timestamps, i.e. the last date and time a user had

shown activity in a given community, so that the past historical data can be used for training and the recent history for testing. For each user, the first 80% of the samples of each class are selected to be used for training, while the remaining 20% were used for testing.

We considered three type of classifiers for the experiments, logistic regression, a neural network and a random forest classifier. An optimal classifier, one that performs the best on the validation dataset, is selected for each of the three models. For the decision-based fusion model, it is an ensemble of classifiers, since each model employs a different one i.e. the best-performing classifier.

4   Discussion of Results

A set of six evaluation metrics, namely, accuracy, precision, recall, F1, area under the ROC curve and cross-entropy loss was used to validate and compare the effectiveness of the proposed models. The evaluation results from the testing experiments are summarized in Table 1.

Table 1. Evaluation results for each of the models

Model	Classifier	Accuracy	Precision	Recall	F1	ROC AUC	Cross-entropy loss
Network statistics	Neural Network	0.8631	0.8764	0.8631	0.8619	0.8631	0.2824
Node embeddings	Random Forest	0.8529	0.8628	0.8529	0.8519	0.8529	0.3156
Content analysis	Random Forest	0.8432	0.8553	0.8432	0.8419	0.8432	0.3558
Fusion - Majority	Ensemble	0.8615	0.8721	0.8615	0.8605	0.9881	0.1878
Fusion - All features	Random Forest	0.9269	0.9286	0.9269	0.9268	0.9862	0.1795

The three models that are based on one type of features, network-related only, network embeddings only, and content analysis only, have shown comparable results across all performance metrics, with the first model yielding the best results. We should note, that while superior in the results, the first model is computationally-demanding, frequent updates might prove too expensive and timely, and is more biased toward the familiar or popular choices. The other two models based on network embeddings and content similarity, seem like a promising and more robust choices, with the latter one especially suited for discovering hidden and new user's interests.

The fusion-ensemble model displays comparable results in all metrics, except a 13% increase of ROC AUC and a 10% drop in cross-entropy loss. The ROC curve plots for the models's performance displayed in Figs. 5 and 6, show that the two multi-source fusion model show comparable performance and that are not greatly affected by the drop in accuracy. Close inspection of misclassified data gave us some insights into the possible reason for the slightly lower performance yielded by the fusion-ensemble model. Namely, the problem lies in the fact that the content similarity model frequently disagrees with the network based models.

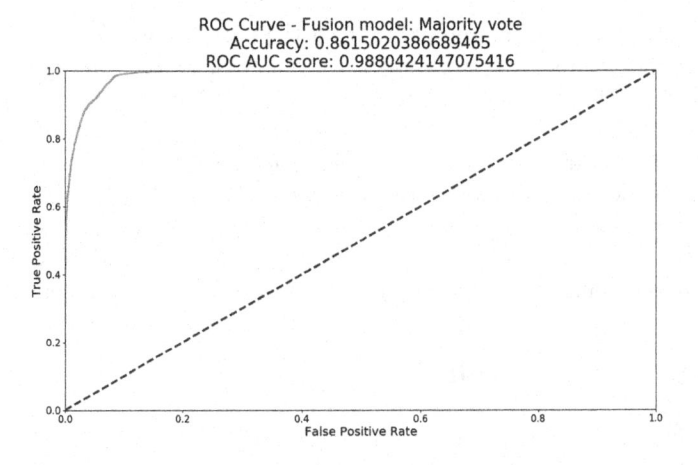

Fig. 5. ROC curve for the majority-based fusion model.

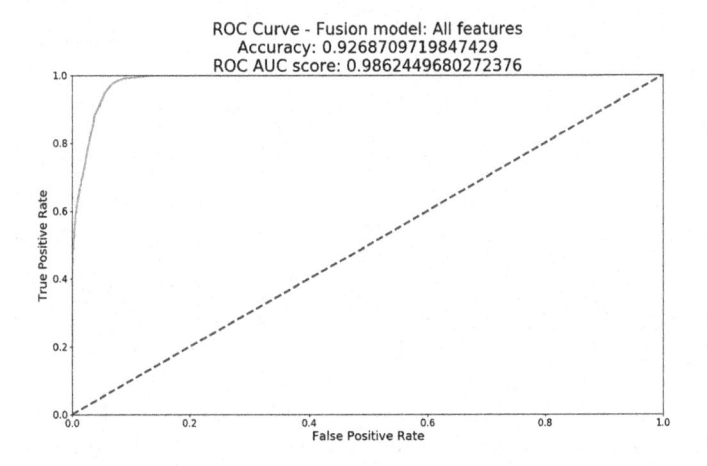

Fig. 6. ROC curve for the fusion model trained on all features

This research has highlighted the relevance of both, network and content linguistic features when fusing different sources of information. However, the analysis has suggested several directions that can be pursued to extend the current research. One research question that could be particularly useful investigating is a strategy for selecting the best model based on the particularities of the situation at hand. Depending on the outcome of interest one wishes to predict (e.g., recommendation for a new user v.s. "surprise me" recommendation), one may wish to employ the most effective model which is best suited for the situation.

The fusion-ensemble model falls under the category of decision-based fusions, currently fusing by majority voting of the estimates of the three models, however the decision function is not restricted to any method per se as other strategies (e.g., weighted, best) can be plugged in and further investigated. Our current

research efforts are directed toward more experimentations, as we continue to refine the diverse set of features included to model users' interests for certain topics i.e., subreddits.

The relevance of any study can only be established at the juxtaposition with previous research that made a significant impact in the field. There are few research studies exploring objectives related to our research. Comparative performance analysis of the results of our study with other approaches is challenged as usual by the differences in the quality and quantity of the datasets, context of use and the objectives of our research, which is recommending subreddit groups a user should join. The diverse set of features included in the models we propose and the encouraging results we have reported warrant further research efforts to strengthen the work.

5   Conclusions

In this paper, several models for subreddit recommendations have been proposed and evaluated. As the basis for the models, two user-subreddit interaction networks were created that capture the knowledge on users' activity within the themed subreddit communities. Creating value from social networks cannot be discussed without regard to the ongoing advances and challenges in related fields, from graph analysis to social data mining and behavioral analytics to natural language processing.

Drawing on successful practices and experiences of related studies, a selection of features, diverse by nature (network vs. linguistic features) has been investigated, and two fusion models were proposed to accommodate much broader recommendation scenarios. The analysis points out to the role each type of features could play in recommending communities to Reddit users in tune with their interests and preferences. The main challenge for future research would be to select the most suitable fusion method i.e., a strategy that recognizes the strengths and weaknesses of each type of modelling feature set to reach the ultimate goal of devising a successful recommender system.

Acknowledgement. This work was partially financed by the Faculty of Computer Science and Engineering at the "Ss. Cyril and Methodius" University.

References

1. Bird, S., Klein, E., Loper, E.: Natural language processing with Python. O'Reilly Media (2009). http://www.datascienceassn.org/sites/default/files/Natural%20Language%20Processing%20with%20Python.pdf
2. Clauset, A., Newman, M.E., Moore, C.: Finding community structure in very large networks. Phys. Rev. E **70**(6), 066111 (2004). https://doi.org/10.1103/PhysRevE.70.066111
3. Das, A.K., Bhat, N., Guha, S., Palan, J.: A personalized subreddit recommendation engine. CoRR abs/1905.01263 (2019). https://arxiv.org/pdf/1905.01263.pdf

4. Grover, A., Leskovec, J.: node2vec: scalable feature learning for networks. In: Proceedings of the 22nd ACM SIGKDD International Conference on Knowledge Discovery and Data Mining, pp. 855–864. ACM (2016). https://doi.org/10.1145/2939672.2939754

5. Hagberg, A.A., Schult, D.A., Swart, P.J.: Exploring network structure, dynamics, and function using networkx. In: Varoquaux, G., Vaught, T., Millman, J. (eds.) Proceedings of the 7th Python in Science Conference, Pasadena, CA USA, pp. 11–15 (2008). https://www.osti.gov/biblio/960616

6. Huang, R., McIntyre, S., Song, M., Ou, Z., et al.: An attention-based recommender system to predict contextual intent based on choice histories across and within sessions. Appl. Sci. **8**(12), 2426 (2018). https://doi.org/10.3390/app8122426

7. Jamonnak, S., Kilgallin, J., Chan, C.C., Cheng, E.: Recommenddit: a recommendation service for reddit communities. In: 2015 International Conference on Computational Science and Computational Intelligence (CSCI), pp. 374–379. IEEE (2015). https://doi.org/10.1109/CSCI.2015.64

8. Lakkaraju, H., McAuley, J., Leskovec, J.: What's in a name? Understanding the interplay between titles, content, and communities in social media. In: Seventh International AAAI Conference on Weblogs and Social Media (2013). https://www.aaai.org/ocs/index.php/ICWSM/ICWSM13/paper/viewPaper/6085

9. MacLean, C., Garza, B., Oganesian, S.: A recurrent neural network based subreddit recommendation system (2016). http://cole-maclean.github.io/blog/files/subreddit-recommender.pdf. Accessed 12 Sept 2019

10. Mikolov, T., Sutskever, I., Chen, K., Corrado, G.S., Dean, J.: Distributed representations of words and phrases and their compositionality. In: Burges, C.J.C., Bottou, L., Welling, M., Ghahramani, Z., Weinberger, K.Q. (eds.) Advances in Neural Information Processing Systems, vol. 26, pp. 3111–3119. Curran Associates, Inc. (2013). http://papers.nips.cc/paper/5021-distributed-representations-of-words-and-phrases-and-their-compositionality.pdf

11. Nguyen, H., Richards, R., Chan, C.C., Liszka, K.J.: RedTweet: recommendation engine for reddit. J. Intell. Inf. Syst. **47**(2), 247–265 (2016). https://doi.org/10.1007/s10844-016-0410-y

12. Pedregosa, F., et al.: Scikit-learn: machine learning in Python. J. Mach. Learn. Res. **12**, 2825–2830 (2011). http://www.jmlr.org/papers/volume12/pedregosa11a/pedregosa11a.pdf

13. Poon, D., Wu, Y., Zhang, D.Q.: Reddit recommendation system (2011). https://pdfs.semanticscholar.org/f8ae/316fce59e0ca585aaf85e8f18a796134fc5d.pdf

14. Sundaresan, V., Hsu, I., Chang, D.: Subreddit recommendations within Reddit communities (2014). http://snap.stanford.edu/class/cs224w-2014/projects/cs224w-16-final.pdf. Accessed 12 Sept 2019

15. Tuomchomtam, S., Soonthornphisaj, N.: Community recommendation for text post in social media: a case study on Reddit. Intell. Data Anal. **23**(2), 407–424 (2019). https://doi.org/10.3233/IDA-183861

Application of Hierarchical Bayesian Model in Ophtalmological Study

Biljana Tojtovska[1(✉)], Panche Ribarski[1], and Antonela Ljubic[2]

[1] Faculty of Computer Science and Engineering, Ss. Cyril and Methodius University, Skopje, Macedonia
{biljana.tojtovska,pance.ribarski}@finki.ukim.mk
[2] Private Polyclinic "Medika Plus", Skopje, Macedonia
ljubicantonela@gmail.com

Abstract. The problems with statistical results based on p-values, together with multiple comparisons have been criticized often in the literature. Many authors argue that this way of reporting scientific research creates unreliable results. This issue is especially important in the era of Big Data, when many tests are done on the same data sets, which are often openly available. A way to overcome these problems is offered by Bayesian analysis. In our previous research we have used traditional statistical approach to conduct multiple hypothesis tests on our data in ophtalmological study. The goal of this paper is to apply the hierarchical Poisson exponential model on the data and test the dependence of congenital heart disease and Brusfield spots. We give detailed description of the model, analyze the generated Markov chains and the posterior distributions for the simulated parameters and discuss the results from Bayesian perspective. The results are original and have not been published yet.

Keywords: Bayesian methods for test of independence · Contingency table · Hierarchical Poisson exponential model

1 Introduction

In the recent decades, many scientists have criticized some practices in the scientific research. This, among the other, includes bad practices in study design, unreproducible studies, small study power and the continuous chase for significant results [6,12,18,23]. p-values have become a publishing criteria in many areas, which has lead to the so called "drawer effect" and the problem of p-hacking [11,17,20,25]. In [21] Simmons et al. give guidelines for both authors and researchers in the hope to establish a scientific practice that would slowly

Partially supported by Faculty of Computer Science and Engineering at the University Ss. Cyril and Methodius in Skopje, Macedonia, as part of the project "Stability of coupled stochastic complex networks".

© Springer Nature Switzerland AG 2019
S. Gievska and G. Madjarov (Eds.): ICT Innovations 2019, CCIS 1110, pp. 109–120, 2019.
https://doi.org/10.1007/978-3-030-33110-8_10

overcome the presented problems. However, some authors point out other important problems in the classical framework of hypothesis testing. Kruschke in [14] explains how the p-values and confidence intervals depend upon the researcher's intentions. The researcher may decide upfront to collect a sample with fixed size, but he may also decide to run the experiments for a fixed amount of time or until some condition is fulfilled. Though the sample size may be the same in all three cases, the experimental design is not. The sampling plan determines the underlying probability distribution and thus may considerably influence p-values, confidence intervals and hence data interpretation and conclusions. In [10] Gelman addresses the problem of multiple hypothesis testing. In many exploratory studies in new research fields, but also in fields like genomics and neural imaging, it is often necessary to test the relationship of large number of variables. This increases the family wise error rate, i.e. the probability of false positive results (false discovery). Many adjustment techniques have been suggested (see [24] and references therein) but this does not completely solve the issue. Many of these methods may be conservative, there is often no criteria that recommends their application and even more, their unclear reporting may create results which are hard to compare. What is more important, as Gelman argues, when applying these adjustment techniques we have to consider all possible comparisons that could be reported in the data. This includes all current and future possible research on the same data set. This becomes especially relevant in the era of Big Data, when many data sets are openly available for different analysis, but only partial results are reported as significant findings.

As a solution, both Kruschke and Gelman suggest applying Bayesian methods, to overcome the problem with p-values and multiple comparisons. In Bayesian data analysis the results are not reported in terms of p-values. Instead, based on the posterior distribution we can decide which parameter values are most credible and we can define measures of uncertainty based directly on posterior credible intervals. For details on Bayesian data analysis we refer to [9,13]. However this approach is still not widely accepted, especially in medical sciences, where traditional statistical approach is still a standard and the research articles are interpreted based only on p-values. One of the reasons may be the fact that the Bayesian methods rely heavily on simulation algorithms for the posterior distribution, generally referred to as Markov chain Monte Carlo (MCMC). In the recent years different software became available, which automatically creates samplers for the models (ex. BUGS, JAGS, STAN). This gave the Bayesian methods more practical use and popularity.

The previous discussion has motivated us to extend the statistical analysis of our previously published ophtalmological study [16] and get additional information by conducting some post-hoc Bayesian analysis. The goal of this paper is to apply the hierarchical Poisson exponential model for independence on our data and discuss the methods and results. To the best of our knowledge, Bayesian methods have been rarely applied in the ophtalmological research and there are no applications on our population. Thus we have no information that can be included in our prior but we believe the results can be informative for some future research.

The rest of the paper is structured as follows - in Sect. 2 we give details on our ophtalmological study, the previously conducted analysis and results, and motivate our post-hoc analysis. In Sect. 3 we refer to different Bayesian approaches for tests of independence, and give details on the hierarchical Poisson exponential model. The model is applied to our data and we discuss the results of the simulation. Finally we draw conclusions in Sect. 4.

2   Study on Ophthalmic Manifestation in Children and Young Adults with Down Syndrome and Congenital Heart Disease

We have conducted a study which included 185 children and young adults from Caucasian population with Down Syndrom (DS), who reported presence (51 subject) or absence (134 subjects) of congenital heart disease (CHD). The goal of the study was to investigate association of CHD with different types of ocular manifestations and to compare the results with studies based on data from other regions in the world. In the study we tested 30 hypothesis using Welch two-sided t-test, two-tailed Student's test, Levene's test for equality of variance, χ^2 or Fisher's exact test of independence. We reported mean differences or odds ratios, together with confidence intervals, supported by p-values. The results were published in [16].

We did not apply any correction for multiple comparisons, for few reasons. Many of the correction methods are conservative, and the adjustments may lead to rejection of findings. In [24] we have compared different adjustment methods and we concluded that all the adjusted p-values turned out to be not significant. However, the comparisons do give an insight in this exploratory study and may give direction for future research on the topic. Additionally, there are only few results in the literature on this topic - in [1, 3, 4] the authors consider different ocular manifestations and systemic diseases in DS patients from different populations and only [22] gives results based on a Caucasian population which is geographically close to ours. These studies did not report multiple comparison corrections, which enables us to easily compare the results between studies. However, any further post-hoc analysis should apply adjustments. Motivated by the previous discussion, we have decided to analyze the data using Bayesian methods. This will lead to more detailed inference, since the posterior distribution is more informative and we can also analyze joint probabilities of combination of parameter values.

Brusfield spots are collagenous tissue that may be present in the iris, with a higher frequency in the DS population. In [5] the authors report association between heart defects in DS population and genetic variations of a gene which encodes type of a collagen. However, there is no definitely established medical association. The observed frequencies from our study are presented in Table 1. We reported that the presence of Brushfield spots is not independent from CHD (χ^2 test of independence, p = 0.03). We concluded that additional studies are needed to confirm statistical and explain medical significance.

Table 1. Observed frequency for different levels of CHD and BS in the DS population of children and young adults.

$x_2 \setminus x_1$	With CHD	Without CHD	Total
With BS	15	21	36
Without BS	33	110	143
Total	48	179	183

The goal of this paper is to use Bayesian approach for the test of independence for the data in Table 1 and discuss the results from Bayesian perspective. In the next section we explain in more details the hierarchical Poisson exponential model which will be applied to the data.

3   Hierarchical Poisson Exponential Model, Results and Discussion

In [2] the author describes different Bayesian methods, which can be applied to estimating probabilities of categories, testing independence in two-way contingency tables and calculating Bayes factor. It also includes Bayesian interaction analyses and analyses of general linear models. In [19] the authors give details on Bayesian inference on contingency tables based on Poisson ANOVA model. The hierarchical ANOVA model which we use was popularized by Gelman [7,8]. Next we present the model following the second edition of the book of Kruschke [13].

Let us assume that we have two factors of influence (categories) x_1 and x_2 which are predictor variables, with number of levels n_1 and n_2, respectively. The data is usually represented in a contingency table, where the columns are labeled by the levels of x_1 and the rows are labeled by the levels of x_2, with total of $n_1 \cdot n_2 = n$ cells (ex. Table 1). With $f_{r,c}$ we denote the data (observed frequency) in the (r,c) cell. If the categories are independent, then their joint probability is product of marginal probabilities (multiplicative independence). Thus we need to estimate the marginal probabilities of the categories. Let us denote with $\hat{f}_{rc}$ the predicted frequency for the (r,c) cell, and $\hat{f}_r$ and $\hat{f}_c$ are the predicted marginal frequencies for the r-th row and the c-th column. If N is the sample size (total count), then the independence implies

$$\frac{\hat{f}_{rc}}{N} = \frac{\hat{f}_r}{N} \cdot \frac{\hat{f}_c}{N}$$

This can be transformed to

$$\hat{f}_{r,c} = \exp\left(\ln\frac{1}{N} + \ln\hat{f}_r + \ln\hat{f}_c\right).$$

Hence we can model the cell tendency $\lambda_{r,c}$ as following

$$\lambda_{r,c} = \exp\left(\beta_0 + \beta_r + \beta_c\right) \tag{1}$$

where $\beta_0 = \ln\frac{1}{N}$ is the overall central tendency (baseline), while $\beta_r = \ln f_r$ and $\beta_c = \ln f_c$ are the drifts away from the baseline, which come from the corresponding factor of influence (also called deflections). As in the case of standard ANOVA, the necessary constraints are given by $\sum_r \beta_r = 0$ and $\sum_c \beta_c = 0$.

The Eq. (1) obeys multiplicative independence, which is the desired property for the contingency table. In order to model the possible dependence, we have to add a term which may violate this property. This is achieved by adding interaction terms, which describe the joint influence of the factors. Thus the Eq. (1) is transformed to

$$\lambda_{r,c} = \exp\left(\beta_0 + \beta_r + \beta_c + \beta_{r,c}\right) \tag{2}$$

where $(\beta_{r,c})_{n_1 \times n_2}$ matrix has rows and columns that sum to zero, i.e. $\sum_r \beta_{r,c} = 0$ for all c and $\sum_c \beta_{r,c} = 0$ for all r. Thus, the analysis of the interaction terms may tell us more about the violation of the independence. Next we should specify a likelihood function which, given the mean $\lambda_{r,c}$, will calculate the probability of the observed frequency $f_{r,c}$. Since in our model the sample size N was not fixed apriori, it is suitable to use the Poisson distribution with $\lambda_{r,c}$ being the average rate of occurrence of a subject which reports both of the characteristics r and c (data belonging to (r, c) cell). The average rates will be calculated based on the observed frequencies $f_{r,c}$.

To easily write the simulation, the data is reordered - the two factors and their labels are presented in columns, such that each row represents one (r, c) combination. Then the count of the (r, c) combination in the $i-th$ row is denoted by y_i (ex. Table 2).

Table 2. Reorganization of Table 1.

x_1	x_2	y
With CHD	With BS	15
With CHD	Without BS	33
Without CHD	With BS	21
Without CHD	Without BS	110

In our model, y_i comes from Poison distribution $\mathcal{P}(\lambda_i)$ where

$$\lambda_i = \exp\left(\beta_0 + \sum_j \beta_{1,j} x_{1,j}(i) + \sum_k \beta_{2,k} x_{2,k}(i) + \sum_{j,k} \beta_{1\times2,(j,k)} x_{1\times2,(j,k)}(i)\right) \tag{3}$$

or in vector form

$$\lambda = \exp\left(\beta_0 + \overrightarrow{\beta_1} \cdot \overrightarrow{x_1} + \overrightarrow{\beta_2} \cdot \overrightarrow{x_2} + \overrightarrow{\beta_{1\times2}} \cdot \overrightarrow{x_{1\times2}}\right) \tag{4}$$

Here $\vec{x_k} = \langle x_{k,1}, ..., x_{k,n_k} \rangle$ for $k = 1, 2$ and when the subject has the level j from the corresponding category, then $x_{k,j} = 1$ and $x_{k,i} = 0$ for $i \neq j$. Similar, $\overrightarrow{x_{1 \times 2}}$ is a matrix which corresponds to $n_1 \times n_2$ combinations of the levels of both predictors. When a subject has a particular combination (j, k), then the (j, k) entry of the matrix equals to 1 and all other entries are zero. The components of $\overrightarrow{\beta_{1 \times 2}}$ are also called interaction coefficients.

The hierarchical model first defines prior distributions for the baseline and the deflection parameters in (3). The choices for the prior may be different, depending on the previous knowledge of the researcher and his belief in the model. Since no similar research on the topic was available to us, we chose the prior presented in [13]. At this stage of the research we did not test and compare the results from different priors.

Based on the defined prior, the MCMC simulates representative parameter values from the posterior distribution, but the simulated deflection parameters will not be centered. Thus we denote the simulated parameters by a_0, a_1, a_2 and $a_{1 \times 2}$ and the deflection parameters are transformed to β_1, β_2 and $\beta_{1 \times 2}$ and they will satisfy the given constraints.

The mean and standard deviation for the parameters are chosen to comply with the logarithm of the data. For the mean of the baseline parameter a_0 we choose $\ln \bar{y} = \ln \frac{\sum_i y_i}{N}$ - this is the case when all the cells in the contingency table have equal counts. The highest standard deviation of the data S' is achieved when all cells, but one, are equal to zero, i.e. all subjects report only one (r, c) combination. We select a broad prior by taking $a_0 \sim \mathcal{N}(\ln \bar{y}, \frac{1}{(S')^2})$. The components of the vectors a_1, a_2 and the matrix $a_{1 \times 2}$ are assumed to be normally distributed with means zero, and standard deviation $\frac{1}{\sigma_\alpha}$. Here σ_α is generated from a gamma distribution with mode $\ln \bar{y}$ and standard deviation $2 \cdot S'$.

The simulation of the parameters was done in R and we used the OpenBUGS version of BUGS (Bayesian inference Using Gibbs Sampling). Our code is an adjustment of the code presented in the first edition of the book of Kruschke [15]. For implementation using JAGS or STAN we refer to the second edition of the book [13].

An important characteristic which shows how much the chains change from step to step is autocorrelation. The first simulations showed very high autocorrelation for all the simulation parameters. This means that it will take a long time for the chain to cover (investigate) the full range of the distribution. Based on this we decided to generate 4 MCMC samples with length 500000 from the posterior distribution of all simulated parameters. However, some regions of the distribution may still be over represented. Thus we use thinning of $m = 500$, meaning we keep every m-th value in the sample. We also need to check whether the generated values are representative of the posterior distribution. The starting point of the chain may influence the convergence, and also the early steps of the chain may be unrepresentative. This initial iterations may be discarded and we have set this so called burn-in steps to 50000. The results of the simulation of the $a_{1 \times 2}$ parameters are given below - Fig. 1 represents the trace plot of the Markov chain trajectories for parameters $a_{1 \times 2}$, after the burn-in steps.

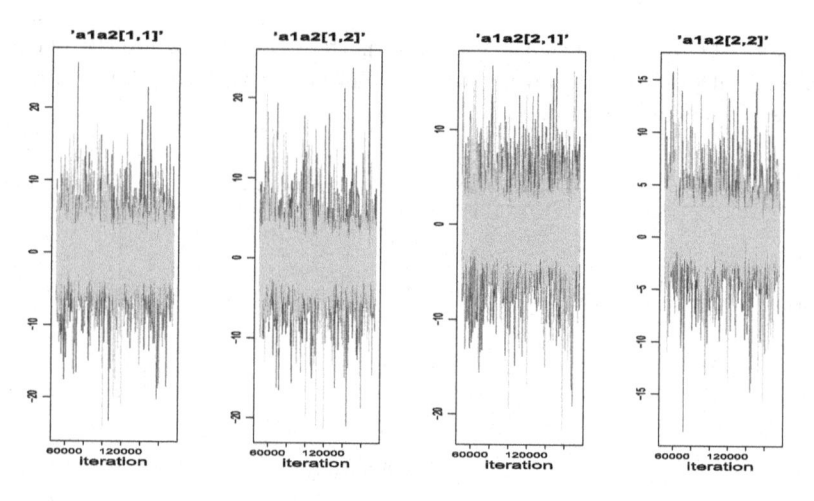

Fig. 1. Trace plots for the Markov chain trajectories for the simulated parameters $a_{1\times2}$, represented after the burn-in period. (Color figure online)

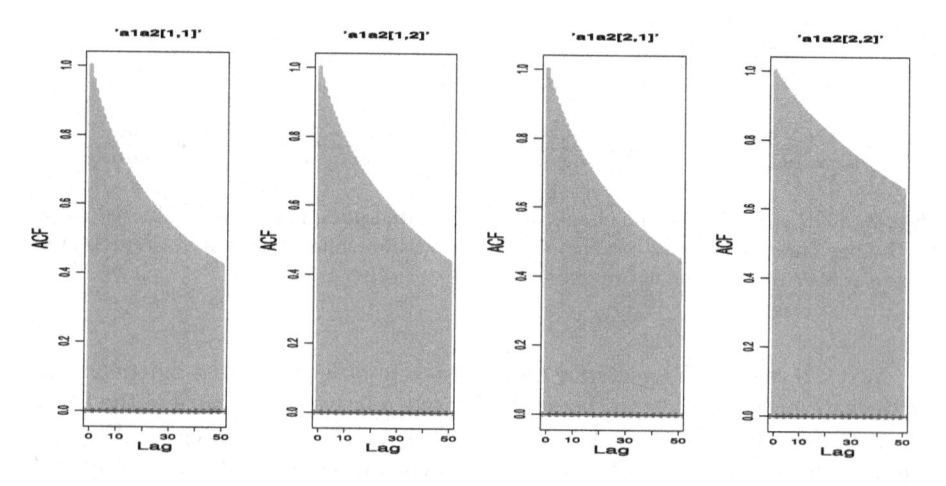

Fig. 2. Autocorrelation function for chains for the simulated parameters $a_{1\times2}$ with lag $L = \overline{1,30}$. (Color figure online)

In Fig. 2 we give autocorrelation of $a_{1\times2}$ chains with lag $L = \overline{1,30}$. The autocorrelation is high and since it may depend on the sampler, we may try to use different sampler (ex. Hamiltonian Monte Carlo).

There are different measures to check whether the chains explore the whole distribution and do not get stuck in an unrepresentative region of the parameter space. For each parameter we checked whether the four chains are well mixed. We used the Brooks-Gelman-Rubin measure, which in BUGS is called *bgr* statistics. It compares the variance between chains with the variance within the chains. If this ratio is close to 1, it shows that the chains are well mixed. Figure 3 shows this

analysis for the $a_{1\times2}$ chains - the red curve is the bgr curve, while the between-chain and within-chain variances are represented in green and blue. Thus, the simulated chains are well mixed.

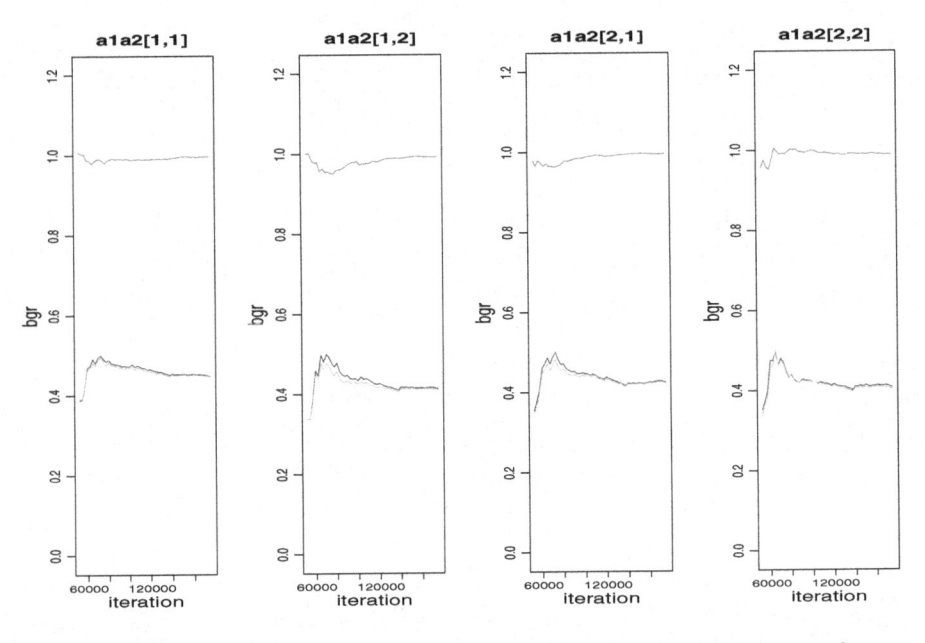

Fig. 3. BGR statistics for the chains of the simulated parameters $a_{1\times2}$ - the red curve is the bgr curve, while the between-chain and within-chain variances are represented in green and blue. (Color figure online)

Next we analyze the posterior distribution. In order to discuss independence of the categories, we have to check whether the relative proportions of the levels of one category are independent on the levels of the other category. For this we have to consider the differences of the corresponding deflections. Figure 4 shows the marginal histograms of the posterior distribution. This figure helps us analyze the histograms for the interaction parameters of each cell.

The histogram corresponding to no presence of CHD and presence of BS shows interaction parameter values below zero - this means that this combination of levels happens less frequently than expected if it was true that the two categories are independent. We can draw similar conclusions from the other histograms. The 95% highest density intervals (HDI) are plotted on each histogram. The HDI contains the values of the distribution which are most credible and cover the 95% of the distribution. The width of the interval is a measure for the degree of certainty - a wide HDI implies uncertainty. We remark that all the HDI for the single-cell interaction coefficients do not include zero, based on which one may conclude that there is dependence between the categories. However, all HDI are very close to 0. In our original research we have used the

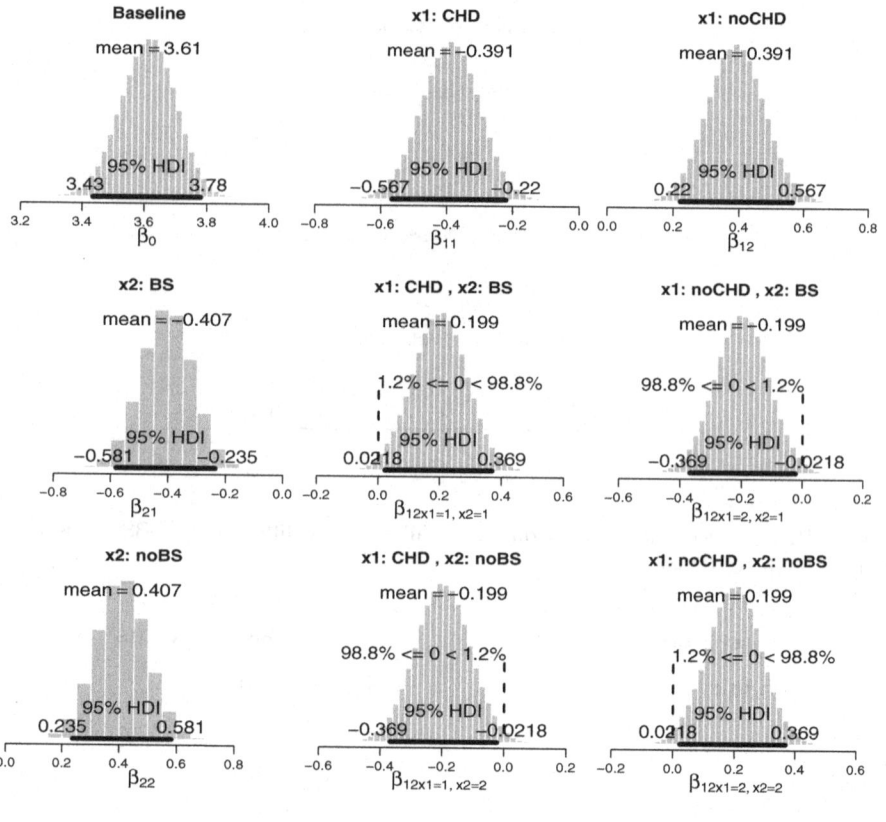

Fig. 4. Posterior distributions for the hierarchical Poisson exponential model applied to the data in Table 1.

χ^2 test to analyze the independence of CHD and Brusfield spots. Based on the p-value only (p = 0.03) we may conclude the two characteristics are dependent. However, if we apply multiple comparison corrections as in [24], we may conclude the contrary. The current analysis indicates there may be independence, but additional information may be needed.

One way to get some additional information is through the analysis of the histograms of different interaction contrasts - this analysis may tell us where does the dependence come from. Figure 5 shows one of the interaction contrasts, difference of differences of BS levels with respect to the CHD levels.

We can see from Table 1 that the difference of BS levels in the first level of CHD category is significantly different from the difference of BS levels in the second level of CHD category. The simulation results in Fig. 5 are presented together with the HDI values which are negative and HDI again does not contain the zero. This interaction contrast analyses gives us more details on the dependence of the characteristics, without the cost of multiple hypothesis testing. Since this is one of the first studies considering these characteristics, new data will update the posterior and will lead to additional information.

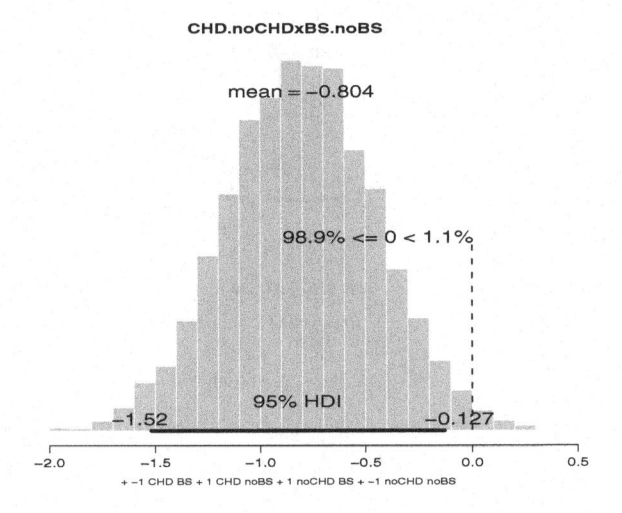

Fig. 5. Histogram of interaction contrast - difference of differences of BS levels with respect to the CHD levels.

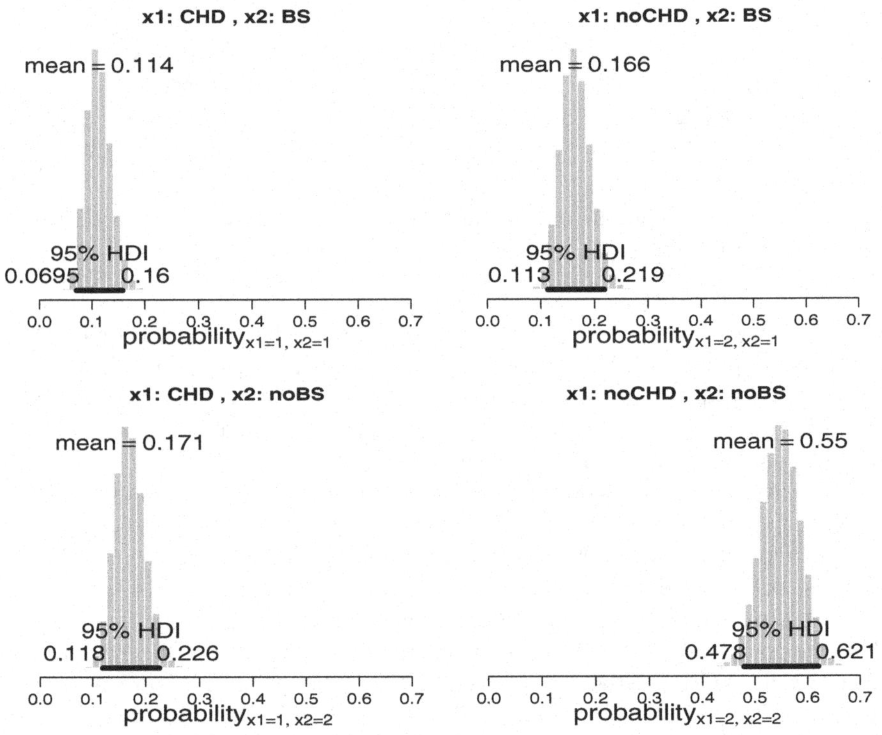

Fig. 6. Distribution of credible cell probabilities. (Remark. The presence and absence of a characteristic are denoted by 1 and 2, respectively.)

Last in Fig. 6 we give the histograms for the estimated cell probabilities for our data. As expected, the probability for absence of both CHD and BS is the highest, which corresponds to the data and the previous contrast analyses.

The presented model can be applied to test independence of any of the factors which were considered in the original study, and this can be extended with any number of posthoc tests. We can try to improve the convergence of the chains with choosing different starting point - for ex. we can start at a point in the middle of the distribution, or at the maximum likelihood of the estimate of the parameter if we assume that the data will overpower the prior. We can compare the results of simulations based on different prior distributions. One can also describe effects of shrinkage, calculate region of practical equivalence (ROPE), effective sample size (ESS), power analysis etc. This is all possible without any concerns of multiple testing, since the number of tests does not affect the interpretation of the results from Bayesian analysis.

4   Conclusions

In this paper we applied the Bayesian hierarchical Poisson exponential model to test the independence of Brushfield spots and congenital heart defects in population of children and young adults with Down syndrome. We have motivated the Bayesian approach by addressing the problems with p-values and multiple comparison in the traditional statistical hypothesis testing. In the model we have defined the prior distributions for the parameters, exponential link function and based on these we generated MCMC samples for the parameters. We analyzed the simulated Markov chains to ensure convergence and we reported details on one of the deflection parameters. We also analyzed the posterior distribution for the parameters and the cell probabilities. Based on the data and the model, the conducted Bayesian analysis indicates that the two features may not be independent. The method gives more detailed information compared to the original χ^2 test. Additional tests can be conducted and the conclusions can be updated with every new data collected.

References

1. Afifi, H.H., Abdel Azeem, A.A., El-Bassyouni, H.T., Gheith, M.E., Rizk, A., Bateman, J.B.: Distinct ocular expression in infants and children with down syndrome in cairo, Egypt: myopia and heart disease. JAMA Ophthalmol. **131**(8), 1057–1066 (2013)
2. Albert, J.H.: Bayesian Methods for Contingency Tables. American Cancer Society (2005)
3. Bromham, N.R., Woodhouse, J.M., Cregg, M., Webb, E., Fraser, W.I.: Heart defects and ocular anomalies in children with Down's syndrome. Br. J. Ophthalmol. **86**(12), 1367–1368 (2002)
4. da Cunha, R.P., de Castro Moreira, J.B.: Ocular findings in Down's Syndrome. Am. J. Ophthalmol. **122**(2), 236–244 (1996)

5. Davies, G.E., et al.: Genetic variation in the COL6A1 region is associated with congenital heart defects in trisomy 21 (Down's Syndrome). Ann. Hum. Genet. **59**, 253–269 (1995)

6. Eklund, A., Nichols, T.E., Knutsson, H.: Cluster failure: why fMRI inferences for spatial extent have inflated false-positive rates. Proc. Natl. Acad. Sci. **113**(28), 7900–7905 (2016)

7. Gelman, A.: Analysis of variance - why it is more important than ever. Ann. Statist. **33**(1), 1–53 (2005)

8. Gelman, A.: Prior distributions for variance parameters in hierarchical models (comment on article by Browne and Draper). Bayesian Anal. **1**(3), 515–534 (2006)

9. Gelman, A., Hill, J.: Data Analysis Using Regression and Multilevel/Hierarchical Models. Analytical Methods for Social Research. Cambridge University Press, Cambridge (2007)

10. Gelman, A., Loken, E.: The statistical crisis in science. Am. Sci. **102**, 460 (2014)

11. Head, M.L., Holman, L., Lanfear, R., Kahn, A.T., Jennions, M.D.: The extent and consequences of p-hacking in science. PLoS Biol. **13**(3), 1–15 (2015)

12. Ioannidis, J.P.A.: Why most published research findings are false. PLOS Med. **2**(8) (2005)

13. Kruschke, J.: Doing Bayesian Data Analysis, 2nd edn. Academic Press, Boston (2015)

14. Kruschke, J.K.: Bayesian data analysis. Wiley interdisciplinary reviews. Cogn. Sci. **15**, 658–676 (2010)

15. Kruschke, J.K.: Doing Bayesian Data Analysis: A Tutorial with R and BUGS, 1st edn. Academic Press Inc., Burlington (2010)

16. Ljubic, A., Trajkovski, V., Tesic, M., Tojtovska, B., Stankovic, B.: Ophthalmic manifestations in children and young adults with Down Syndrome and congenital heart defects. Ophthalmic Epidemiol. **22**(2), 123–129 (2015)

17. Nuzzo, R.: Scientific method: statistical errors. Nature **506**(7487), 150–152 (2014)

18. OSF: Estimating the reproducibility of psychological science. Science **349**(6251) (2015)

19. Robert, C., Marin, J.M.: Bayesian Core: A Practical Approach to Computational Bayesian Statistics. Springer, New York (2007). https://doi.org/10.1007/978-0-387-38983-7

20. Rosenthal, R.: The file drawer problem and tolerance for null results. Psychol. Bull. **86**(3), 638–641 (1979)

21. Simmons, J.P., Nelson, L.D., Simonsohn, U.: False-positive psychology: undisclosed flexibility in data collection and analysis allows presenting anything as significant. Psychol. Sci. **22**(11), 1359–1366 (2011)

22. Stirn Kranjc, B.: Ocular abnormalities and systemic disease in Down Syndrome. Strabismus **20**, 74–77 (2012)

23. Szucs, D., Ioannidis, J.P.: Empirical assessment of published effect sizes and power in the recent cognitive neuroscience and psychology literature. bioRxiv (2016)

24. Tojtovska, B.: Multiple hypothesis testing: adjustment methods with application. In: Proceedings of CIIT 2019–16th International Conference on Informatics and Information Technologies, Mavrovo, N. Macedonia (2019)

25. Wagenmakers, E.J.: A practical solution to the pervasive problems of p-values. Psychon. Bull. Rev. **14**(5), 779–804 (2007)

Friendship Paradox and Hashtag Embedding in the Instagram Social Network

David Serafimov, Miroslav Mirchev[(✉)], and Igor Mishkovski

Faculty of Computer Science and Engineering,
Ss. Cyril and Methodius University in Skopje, Skopje, North Macedonia
miroslav.mirchev@finki.ukim.mk

Abstract. Instagram is a social networking platform which gained popularity even faster than most of the other modern online social networks. It is relatively newer and less explored than other social networks, such as Facebook and Twitter. Therefore, we have conducted a research based on a sample data set extracted through the Instagram weekend hashtag project, in order to unveil some of its characteristics. First, we reveal the various forms of friendship paradox present in Instagram, which are often observed in social networks. Then, we conduct a detailed hashtag analysis and provide a method for hashtag representation and recommendation using natural language processing.

Keywords: Online social networks · Network science · Natural language processing

1 Introduction

Online social networks (OSNs) have been widely studied in the past [11,14,15], however, Instagram is relatively newer and less researched. Knowledge discovery from this network can help gain a deeper insight into the processes that drive its growth, as well as reveal some characteristics of other social networks in the real world for which there is no available data. A thorough study of Instagram was presented in [9] where users and photos were divided into several categories based on network and photographic data. In [10] the authors provide a detailed analysis of the liking activity among the Instagram users. The behavior of the silent users, known as lurkers, was explored in [18] for several OSNs including Instagram. In this paper we focus on two topics, revealing the existence of the friendship paradox and providing a suitable hashtag embedding.

The friendship paradox is a phenomenon discovered in 1991 [5] and it states that "most people have fewer friends than their friends have, on average". On the contrary, usually people think that they have more friends than most of their friends. This phenomenon is not limited to friends and can be observed in social networks with other types of relationships. An example of this is the social

S. Gievska and G. Madjarov (Eds.): ICT Innovations 2019, CCIS 1110, pp. 121–133, 2019.
https://doi.org/10.1007/978-3-030-33110-8_11

network of partners. Most of the individuals in this network have fewer partners than their partners on average. The friendship paradox can be also applied in predicting epidemic spreading as well as immunization [3]. In addition to the real world, this paradox is also present in the online world. One example is the social network Twitter [8]. In this social network, more than 98% of users had fewer followers than their followers, on average. The friendship paradox have been explored in other social networks such as Facebook [7], but to the best of our knowledge it has not been confirmed for Instagram. Here we will show the friendship paradox using a dataset extracted from Instagram, which was already explored in [6] for studying other relevant interesting aspects.

We will use the same data set to provide a hashtag analysis in Instagram using natural language processing, which to the best of our knowledge have not be done elsewhere. Several approaches have been published to get multidimensional representations of hashtags. These methods depend on additional features like images [19], text [4,20], or some other. The nature of these methods does not allow us to easily adapt them for data sets where such features are not available. On the other hand, in our study we rely only on the available hashtags. These network analyses could be useful in exploring the spread of trends across the network [21], and potentially their control and timely prevention.

The paper is organized in the following way. In Sect. 2 we describe the Instagram data set used in our study. In Sect. 3 we explain the friendship paradox and present our analysis for Instagram. In Sect. 4 we show a method for hashtag representation and based on it two models for hashtag recommendation. We finalize the paper with some conclusions in Sect. 5.

2   Data Set

In this paper we study the social network Instagram through a data set based on the weekend hashtag project, which was first used and described in [6]. The weekend hashtag project is a competition that is held every Friday and is organized by the Instagram team. The contest consists of a unique hashtag with the #whp prefix, which is also the theme of the contest. Users of the social network can participate in the competition if they post a picture with the designated hashtag. To collect this data set, 72 WHP hashtags were selected. 2081 users who joined one of these 72 competitions, were randomly selected. Data about what these users shared was also collected. A breadth first search is started from the seed users, skipping any users who did not participate in the contests.

The data set consists of two files. The first file in each row contains: follower id, followed user id, number of likes from the follower to the followed, number of comments from the follower to the followed, and timestamps for all the comments. The second file consists of data about what users posted, where each row contains: posted picture id, number of shares for the post, post timestamp, hashtags included in the post, the number of likes for the post, and the number of comments for the picture. In the data set we have a total of 1, 686, 349 posts. These posts were posted by 2, 081 different users. A total of 8, 919, 630 hashtags

were included in all posts, from which 269, 359 were unique. The total number of likes was 1, 242, 923, 022 and the number of comments was 41, 341, 783. There are a total number of 44, 766 users. There are 677, 686 connections between these users. The average node degree in the network is 15.14. The average length of the shortest path in this network is 3.16, although the longest shortest path is 11. The network has 151 communities. The clustering coefficient is 0.041, the assortativity coefficient is −0.097, and the modularity of the network is 0.578.

3   Friendship Paradox in Instagram

In the Instagram social network, users can do several activities such as follow users, post images or videos, like, comment, etc. In this section we examine whether the friendship paradox applies to this social network and for which activities. The relationships in Instagram are directed, so if we follow someone, they do not have to follow us. The people that follow us are our followers and those that we follow are our followees. Therefore, we check if a friendship paradox occurs both relating to followers and followees in Instagram. This kind of directed relationships are similar to Twitter for which the various forms of friendship paradox have been explored in [8]. On the other hand, in Facebook the relationships are mutual and the graph is undirected, allowing the application of the classical friendship paradox [7].

The friendship paradox can be rephrased in the two following ways:

(i) Our *followees* have or do something more than us on average.
(ii) Our *followers* have or do something more than us on average.

We also check the presence of both weak and strong friendship paradox, where the *weak* paradox is calculated compared to the average and the *strong* is compared to the median in order to reduce the impact of the extremes. In the following text we present the calculated friendship paradox for different activities and the results are summarized in Fig. 1 for the paradox relating to the followees and in Fig. 2 for the paradox relating to the followers. In the text we will mostly comment the results for the weak paradox, while the results for the strong paradox usually follow a similar pattern and the reader can see them in the figures. We only comment the strong paradox in the number of hashtags, because it does not always apply there.

Followers. Following a user in Instagram allows you to see what they post. By analyzing the data we confirmed a friendship paradox for the number of followers that both our followees and followers have. On average 91.71% of the users are followed by fewer users than their followees, while 73.64% of the users have less followers than their followers.

Followees. The friendship paradox was also observed in the number of followees, so the majority of users follow less people than their followers and followees. On average 82.21% of the users follow less users than their followees, and 76.99% of the users have less followees than their followers.

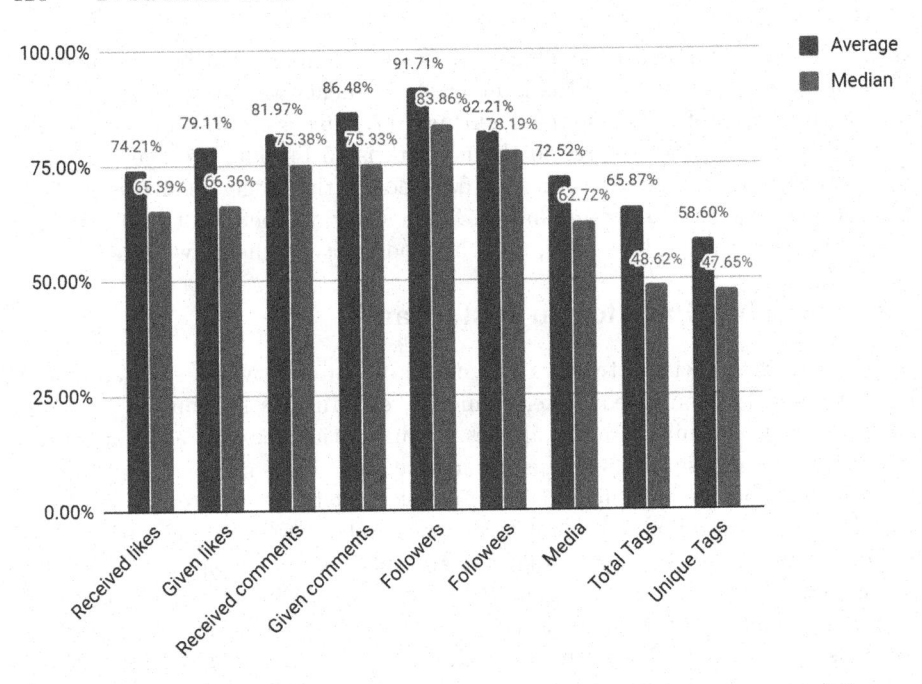

Fig. 1. User percentage for which the friendship paradox relating to the followees applies for various properties.

Likes. Users in Instagram can share likes on posts or comments, and thus let other users know that they like something and the friendship paradox was also shown for the number of likes. On average 74.21% of the users have received less total likes than their followees, while 71.44% of the users have received fewer likes than their followers. The paradox of friendship is also true in the number of likes given. On average 79.11% of users have given less likes than their followees, and 77.57% than their followers.

Comments. In addition to likes, Instagram users can also share text as comments on posts or other activities. With the help of comments, users can share their opinions, discuss publicly on a topic, etc. On average 86.48% of the users have given fewer comments than their followees, while 76.78% have posted less comments from their followers. Similarly, 81.97% of the users have received less comments on their posts than their followees on average, and 69.86% than their followers.

Posts and Hashtags. Instagram users can post and share media. When sharing, users often add hashtags. A hashtag is a string of characters that starts with a hash ('#'). There are several benefits to adding hashtags, such as giving context to a shared image or easier discovery of relevant content. On average 72.52% of the users share less images and videos than their followees, and 68.29%

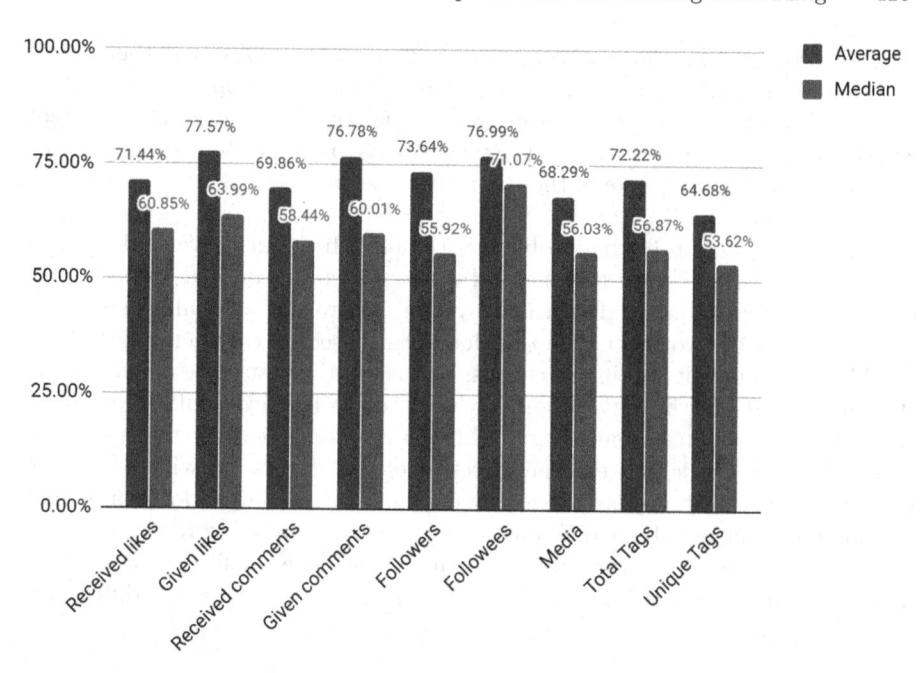

Fig. 2. User percentage for which the friendship paradox relating to the followers applies for various properties.

than their followers. In addition, on average 65.87% of the users used less hashtags per photo than their followees, and 72.22% than their followers. If we consider only unique hashtags, then on average 58.60% of the users have used less unique hashtags per photo than their followees on average, and 64.68% than their followers.

We can conclude that in Instagram's social network, as in other online social networks, we can observe a friendship paradox for various network activities. The weak variant of the friendship paradox applies to all network criteria, while the strong variant of the friendship paradox applies to all criteria except for the number of total and unique hashtags used compared to our followees. For most of the activities the number of followees who have or do something more than us is higher than the number of followers, which is expected, except for the number of hashtags. A higher number of our followers have both total and unique hashtags than our followees. This analysis can help us to better evaluate the impressions and behavior of Instagram users and understand the spreading trends in the social network.

4   Hashtag Analysis with Natural Language Processing

Besides examining the friendship paradox we also conducted an analysis of the hashtags used in Instagram using natural language processing (NLP). NLP is a

discipline concerned with the understanding of natural languages and their representation in machines. For example, the words "dog" and "puppy" have similar semantic meaning, but are composed from different letters and have different lengths. One of the problems that is explored in NLP is how to represent words in such a way that we preserve these semantic meanings, without depending on human intervention.

A popular solution to this problem is the term frequency-inverse document frequency method [17]. For this method we need a corpus of N documents. We represent a word as an N dimensional vector, where the i-th value represents how important the word is to the i-th document based on the occurrence of the word in the document and in the corpus. Let us say the first three documents in the corpus deal with biology. Then words related to biology would have higher values in the first three dimensions, and lower values in the other dimensions. If we calculate the angle between the direction of the vectors, we will get smaller value for words that appear in similar documents, and larger for words that appear in documents about different topics. This method of word presentation has some drawbacks. Vectors are sparse and long. The quality of the vectors depends greatly on the corpus, and it is challenging to preserve subtle differences between similar words.

4.1 Word2Vec

A more sophisticated method for solving this problem is using Word2Vec vectors [13]. This method works using neural networks and a corpus of documents. There are two main approaches. The first approach is to train a neural network to predict words based on its surroundings. The second approach is the opposite, training a neural network to predict the surroundings based on a given word. Both approaches have similar performance. The first approach is generally faster, while the second approach is slower, but the resulting vectors are of higher quality for words that rarely appear in the training corpus. Words that appear in similar environments are represented by vectors with lower difference between their angles. Words that appear in different environments are represented by vectors with greater angle difference between them. The obtained vectors are often used as input for other NLP models, but can be used on their own for solving simpler problems. Other similar methods for word representation in multidimensional space were discovered later, but the obtained vectors did not significantly improve [2,16].

4.2 Hashtag Representation

Although hashtags are different from words, they share many properties. Many of the problems that we encounter in NLP also exist for hashtags, for example: named entity recognition, sentiment analysis, etc. Another common problem is how to represent hashtags. Although several approaches exist, often they depend on additional information like images [19] or text [4,20]. In this paper we will describe a method for embedding hashtags in a multidimensional space based

Table 1. Bag of hashtags example

Input	Output
#mycat #cats #catsofig	#cat
#cat #cats #catsofig	#mycat
#cat #mycat #catsofig	#cats
#cat #mycat #cats	#catsofig

on the Word2Vec method. The input of this method are hashtags that have appeared together. Since we do not have any additional constraints, the proposed method is more general and platform agnostic.

4.3 Architecture

The method consists of a neural network with one hidden layer. On the input we have one hot encoding, meaning that each input node is one hashtag. The hidden layer will have as many nodes as the dimensions of the resulting hashtag vectors. On the output layer we also have one hot encoding. As in Word2Vec, we have two approaches and we will show the difference between them using the following set of hashtags {#cat, #mycat, #cats, #catsofig}. In the first approach the input is all hashtags but one, and the output is the left out hashtag. From one sample with N hashtags we get N inputs and outputs. All the inputs and outputs of the example can be found in Table 1. This approach will be refered to as "bag of hashtags" in this paper.

In the second approach the input is one hashtag, and the output is one of the surrounding hashtags. From one sample with N hashtags we get $N \times (N - 1)$ input output pairs. All the inputs and outputs for the sample can be found in Table 2. This approach will be refereed to as "hashtag pairs" in this paper.

There are several hyperparameters in both approaches:

Number of nodes in the hidden layer. The number of nodes in this layer is equal to the dimensions of the vectors that are obtained at the end. In general, more dimensions can store more information about the hashtags.

Epochs of training. The number of epochs corresponds to the number of times each sample will be used for training the model. The number of epochs linearly increases the training time of the model and improves the quality of the resulting vectors.

Minimum number of occurrences of one hashtag. The number of times that a hashtag should appear in the data set to be included in the vocabulary.

4.4 Evaluation

For evaluating the vector quality, we will use the Instagram data set. Input-output pairs are not available, so we either have to try unsupervised learning or

Table 2. Hashtag pairs example

Input	Output
#cat	#mycat
#cat	#cats
#cat	#catsofig
#mycat	#cat
#mycat	#cats
#mycat	#catsofig
#cats	#cat
#cats	#mycat
#cats	#catsofig
#catsofig	#cat
#catsofig	#mycat
#catsofig	#cats

generate a set of input-output pairs. Since we want to quantify the quality of the vectors, we choose the second approach. The data set does not provide good assumptions about additional hashtags that we could recommend, so instead we have to use the existing data to generate recommendations. We assume that if a user has posted a certain hashtag in a post with multiple hashtags, the hashtag is a good recommendation for the remaining hashtags. This is how evaluation looks like for a single post:

Let us assume the user posted a picture with the following hashtags: #cat, #mycat, #cats and #catsofig. Some hashtags are removed for evaluation, for example #cat and #catsofig. During training we only have #mycat and #cats which we use to generate hashtag embedding. During evaluation we will have the input-output pairs given in Table 3. Such samples, although not perfect, should approximate good recommendations.

Table 3. Evaluation input-output pairs

Input	Output (recommendation)
#mycat #cats #catsofig	#cat
#cat #mycat #cats	#catsofig

4.5   Experiment

The experiment consists of training and evaluating six models of the presented architecture with different hyperparameters compared with a baseline model. The baseline model is a simple statistical model that gives recommendations

Table 4. Pairs of hashtags and their similarity calculated with cosine distance

First hashtag	Second hashtag	Similarity
#instagood	#instamood	0.93206483
#christmas	#xmas	0.87076986
#rap	#rnb	0.74523616
#dad	#father	0.74124840
#netflix	#cats	0.24433972
#nofilter	#sanfrancisco	0.17591012
#instagood	#garden	0.10391730

according to previous occurrences of different hashtags. The data set is split into 90% training and 10% test sets. All models were trained for 50 epochs. Hashtags in the vocabulary have occurred at least 3 times in the data set. Three of the models were trained with the bag of hashtags method, and three with the hashtag pairs method. Two models were trained with 64 nodes in the hidden layer, two with 128 nodes in the hidden layer, and two with 256 nodes in the hidden layer. Learning was executed on the Intel Xeon Scalable processor with 3.7 Gigabytes RAM hosted on the Google Cloud Platform. Model learning takes between one and four hours depending on the method and the hyperparameters on the described hardware. The source code is available on github [1]. We will use the recall at K (R@K) metric to measure quality, which is calculated as the average relevant hashtags that are recommended in the top K, for $K \in 1, 2, 3, 5, 10$.

4.6   Usage

With the resulting vectors, we can perform some operations that were previously difficult or not possible. We can search for similar hashtags, calculate the similarity between hashtags, group hashtags by topics, and even do arithmetic operations with hashtags. All examples in the thesis were obtained from a hashtag pairs model with 64 hidden nodes.

Calculating Hashtag Similarity. Several metrics exist for calculating the distance between two vectors, for example: Euclidean distance, Manhattan distance, cosine distance, etc. If we represent two hashtags as vectors, then with one of the above metrics we could calculate the distance or similarity between the hashtags. Vectors are derived from weights in the neural network, so cosine distance makes most sense in this case. Table 4 provides some examples.

Searching Similar Hashtags. If for a hashtag we calculate the cosine distance with all other hashtags for which we have calculated vectors, we can find the most similar hashtags. Table 5 gives a few examples.

Table 5. Hashtags and their closest neighbors according to cosine distance

Data set	Target hashtag			
	#christmas	#vsco	#istanbul	#healthy
1	#christmastree	#vscocam	#igersistanbul	#eatclean
2	#xmas	#vscophile	#feelingistanbul	#cleaneating
3	#santa	#vscofeature	#turkishfollowers	#igfit
4	#ornaments	#vscogram	#turkey	#exercise
5	#carols	#afterlight	#ig_turkey	#eatingclean
6	#christmaslights	#newvscocam	#hayatandanibarettir	#nutrition
7	#presents	#vscofilm	#igersturkey	#dialabreakfast

Clustering Hashtags. The obtained vectors can be clustered and Fig. 3 shows hierarchical clustering using cosine distance metric and average of clusters as linkage criteria.

Arithmetic Operations. We can perform some arithmetic operations, such as addition and subtraction, with the obtained vectors. If we search for the nearest hashtag of the resultant vector, we can find an approximation to the result. Here are some examples:

#helloween − #pumkin + #christmas = #christmastree
#kids − #little_igers + #cats = #catstagram
#sweden − #stockholm + #turkey = #istanbul
#woods − #forest + #city = #buildings

Calculating Post Distance Based on Included Hashtags. Another interesting feature of word vectors is that with their help we can calculate the distance between documents. Word Mover's Distance [12] is a technique based on Earth Mover's Distance with which we can calculate the similarity of two documents based on the similarity of the words appearing in each document. If we adapt this technique to hashtag embedding, we can calculate the similarity between two images.

4.7   Results

A simple way to find recommendations for n hashtags, is to find the closest hashtags to their average. Although this method is naive, the results show that in practice it is much better than our baseline statistical model. In Table 6 we can see the results. The best results for each metric are in bold.

We can observe that the hashtag pairs method generally performs better on the given task. The number of dimensions also impacts the outcome and the highest dimensional models did not performed better on this task. A possible explanation is that the training set is too small to take advantage of the

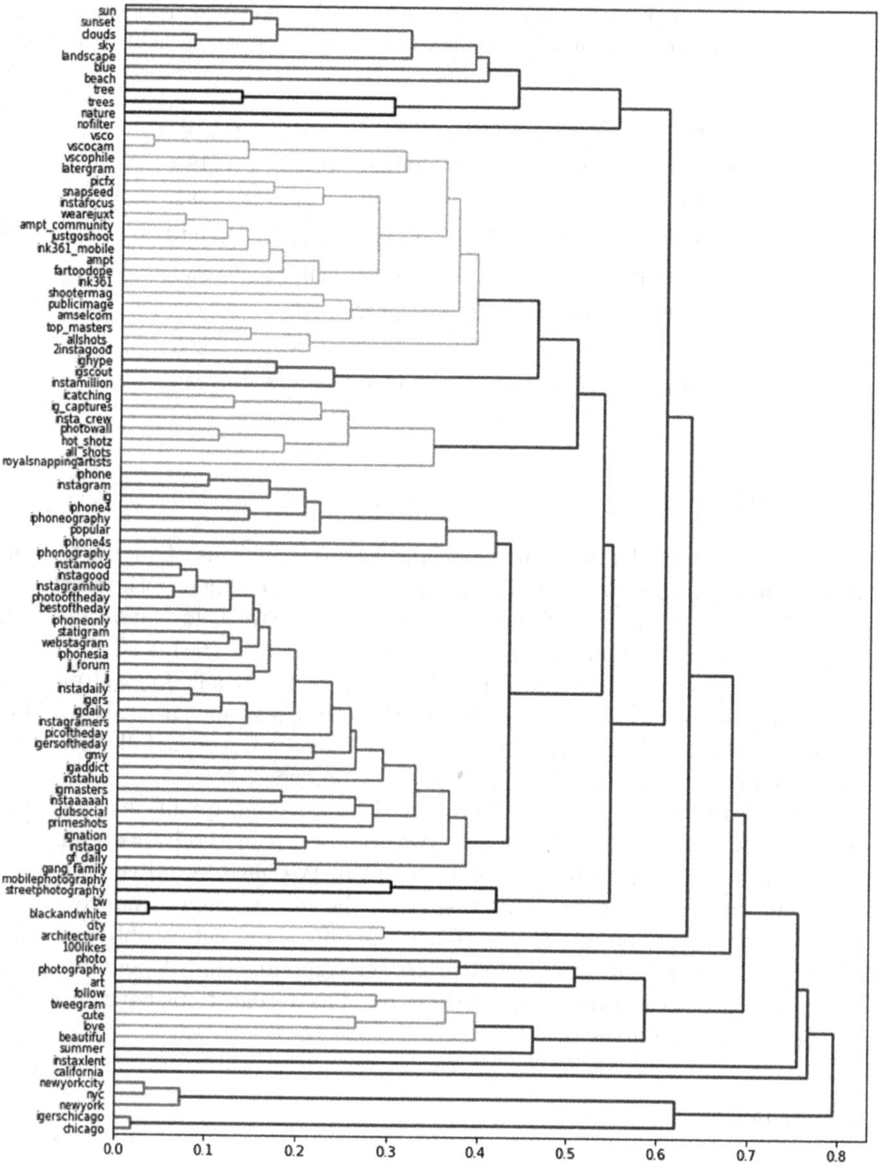

Fig. 3. Hierarchical clustering of the 100 most common hashtags in the dataset.

additional dimensions and the models became overfitted. All models perform on par or better than the baseline model for all metrics. The 128-dimensional model obtained with the hashtag pairs method has the best results for the metrics R@1, R@2 and R@3. The smaller 64-dimensional model performed best for R@5 and R@10 metrics.

Table 6. The recall at K (R@K) metric for the bag of hashtags (BoH), hashtag pairs (HP), with 64, 128 and 256 dimensions (D), as well as the baseline statistical (BS) model.

Model	R@1	R@2	R@3	R@5	R@10
BS	0.0201	0.0366	0.0480	0.0703	0.1184
64D BoH	0.0617	0.0865	0.1035	0.1274	0.1661
64D HP	0.0779	0.1157	0.1435	**0.1836**	**0.2414**
128D BoH	0.0339	0.0477	0.0572	0.0713	0.0956
128D HP	**0.0824**	**0.1189**	**0.1445**	0.1801	0.2307
256D BoH	0.0358	0.0504	0.0623	0.0825	0.1295
256D HP	0.0769	0.1091	0.1296	0.1573	0.1942

5 Conclusion

In this paper we provided some network and hashtag analysis of the Instagram network using a sample data set. In the first part we confirmed both the strong and weak variant of the friendship paradox in the network for many network properties, such as the number of followers, likes, posts, hashtags and comments, both regarding the followers and the followees. Solely for the number of total and unique hashtags compared to the followees, only a weak paradox was observed. Generally the friendship paradox is stronger for the followees than for the followers, except for the number of hashtags used.

We also introduced a general method for obtaining high-quality hashtag representations in multidimensional space. We proposed a method for obtaining a data set for the task of hashtag recommendation. We have tested the obtained hashtag embedding on the given problem and the results showed improvement compared to the baseline model. It is fair to assume that vectors obtained with the proposed models will contribute to improving models that depend on high-quality hashtag representations, similarly as word representations have improved models that depend on them.

References

1. https://github.com/nasadigital/diplomska-instagram
2. Bojanowski, P., Grave, E., Joulin, A., Mikolov, T.: Enriching word vectors with subword information. Trans. Assoc. Comput. Linguist. **5**, 135–146 (2017)
3. Cohen, R., Havlin, S., Ben-Avraham, D.: Efficient immunization strategies for computer networks and populations. Phys. Rev. Lett. **91**(24), 247901 (2003)
4. Dhingra, B., Zhou, Z., Fitzpatrick, D., Muehl, M., Cohen, W.W.: Tweet2Vec: character-based distributed representations for social media. arXiv preprint arXiv:1605.03481 (2016)
5. Feld, S.L.: Why your friends have more friends than you do. Am. J. Sociol. **96**(6), 1464–1477 (1991)

6. Ferrara, E., Interdonato, R., Tagarelli, A.: Online popularity and topical interests through the lens of Instagram. In: Proceedings of the 25th ACM Conference on Hypertext and Social Media, pp. 24–34. ACM (2014)
7. Hampton, K.N., Goulet, L.S., Marlow, C., Rainie, L.: Why most Facebook users get more than they give. Pew Internet Am. Life Proj. **3**, 1–40 (2012)
8. Hodas, N.O., Kooti, F., Lerman, K.: Friendship paradox redux: your friends are more interesting than you. In: Seventh International AAAI Conference on Weblogs and Social Media (2013)
9. Hu, Y., Manikonda, L., Kambhampati, S.: What we Instagram: a first analysis of Instagram photo content and user types. In: Eighth International AAAI Conference on Weblogs and Social Media (2014)
10. Jang, J.Y., Han, K., Lee, D.: No reciprocity in liking photos: analyzing like activities in Instagram. In: Proceedings of the 26th ACM Conference on Hypertext & Social Media, pp. 273–282. ACM (2015)
11. Kumar, R., Novak, J., Tomkins, A.: Structure and evolution of online social networks. In: Yu, P., Han, J., Faloutsos, C. (eds.) Link Mining: Models, Algorithms, and Applications, pp. 337–357. Springer, New York (2010). https://doi.org/10.1007/978-1-4419-6515-8_13
12. Kusner, M., Sun, Y., Kolkin, N., Weinberger, K.: From word embeddings to document distances. In: International Conference on Machine Learning, pp. 957–966 (2015)
13. Mikolov, T., Chen, K., Corrado, G., Dean, J.: Efficient estimation of word representations in vector space. arXiv preprint arXiv:1301.3781 (2013)
14. Mislove, A., Marcon, M., Gummadi, K.P., Druschel, P., Bhattacharjee, B.: Measurement and analysis of online social networks. In: Proceedings of the 7th ACM SIGCOMM Conference on Internet Measurement, pp. 29–42. ACM (2007)
15. Penni, J.: The future of online social networks (OSN): a measurement analysis using social media tools and application. Telematics Inform. **34**(5), 498–517 (2017)
16. Pennington, J., Socher, R., Manning, C.: Glove: global vectors for word representation. In: Proceedings of the 2014 Conference on Empirical Methods in Natural Language Processing (EMNLP), pp. 1532–1543 (2014)
17. Rajaraman, A., Ullman, J.D.: Mining of Massive Datasets. Cambridge University Press, New York (2011)
18. Tagarelli, A., Interdonato, R.: Time-aware analysis and ranking of lurkers in social networks. Soc. Netw. Anal. Min. **5**(1), 46 (2015)
19. Veit, A., Nickel, M., Belongie, S., van der Maaten, L.: Separating self-expression and visual content in hashtag supervision. In: Proceedings of the IEEE Conference on Computer Vision and Pattern Recognition, pp. 5919–5927 (2018)
20. Weston, J., Chopra, S., Adams, K.: #Tagspace: semantic embeddings from hashtags. In: Proceedings of the 2014 Conference on Empirical Methods in Natural Language Processing (EMNLP), pp. 1822–1827 (2014)
21. Zhang, L., Zhao, J., Xu, K.: Who creates trends in online social media: the crowd or opinion leaders? J. Comput. Mediated Commun. **21**(1), 1–16 (2015)

An In-Depth Analysis of Personality Prediction

Filip Despotovski[✉] and Sonja Gievska

Faculty of Computer Science and Engineering, ul.Rudzer Boshkovikj 16, P.O. 393,
1000 Skopje, Republic of North Macedonia
fdespotovski@gmail.com, sonja.gievska@finki.ukim.mk

Abstract. The complex nature of human mind poses recurrent challenges for predictive modeling of human traits and behavior and current technological trends has real potential for advancing the endeavor. In this paper, we present an in-depth analysis of the suitability and effectiveness of several personality prediction models, that incorporate both, multimodal and linguistic features. By studying the impact of various modeling decisions on the predictivness of each Big Five personality dimension, our findings suggest that some modeling choices might be at odds with one another or the objective of the target application scenario, which highlights the importance of extensive experimentation and unique modeling approaches for various aspects of this multi-faceted phenomenon.

Keywords: Personality prediction · Natural language processing · Big Five personality model · Linguistic analysis

1 Introduction

The cognitive processes underlying human behavior are complex and multifaceted and their modeling is still a challenging task for intelligent systems. Our behavior is a mirror of our personality. What we do or participate in, our attitude towards new or our tendency to return to past experiences, our preferences and emotional reactions to certain events or how we carry interactions with others are expressions of our unique character and personality. Analysis of the behavioral manifestations of personality facilitate tangible modeling of the phenomenon within the framework of the existing psychometric models, such as Myers-Briggs Type Indicator [21] or Big Five [11]. For the purpose of this research, we have adopted the Big Five model that describes the personality along five dimensions. There exist wide variations in the behavioral manifestations among the five personality dimensions evident in the listed characteristics below [16]. Discriminating between the personality traits on the opposing ends of each dimension is far from straightforward.

- Extroversion: assertiveness, sociability, talkativeness, tendency to seek stimulation in the company of others

© Springer Nature Switzerland AG 2019
S. Gievska and G. Madjarov (Eds.): ICT Innovations 2019, CCIS 1110, pp. 134–147, 2019.
https://doi.org/10.1007/978-3-030-33110-8_12

- Agreeableness: friendliness, compassionateness, cooperativeness
- Conscientiousness: organized, self-disciplined, efficient
- Neuroticism (Emotional stability): sensitiveness, nervousness, anxiousness
- Openness to experience: inventiveness, curiousness, intellectual, emotional sensitivity, interest in culture, ideas, and aesthetics

As social networks have become a favorite playground for exploration of human individual and collective behavior on a large scale, affective analysis including personality prediction have become the subject of intensive research. A unifying element across these research efforts has been a joint utilization of natural language processing and machine learning methods as crucial discriminators between the effectiveness of the proposed models [13,20,25]. Despite the constant emergence of new research heralding the personality expressions in language use, establishing a clear picture of the causality between language use and personality needs further attention. Rigorous testing and in-depth analysis is what we regard as crucial for confronting the complexity of the problem.

For the purpose of this study, we have chosen the IDIAP YouTube dataset [3,4] consisting of 404 transcribed video monologues (vlogs). Each video is associated with a number of audio-visual features, the gender of the vlogger, and the impressions of personality traits solicited by crowdsourcing. A joint utilization of natural language processing and machine learning, which we chose to undertake follows the line of work of related research in the field and further contributes to the discussion. In particular, in most of the personality prediction studies, once the best performing classifier is selected during initial testing and validation, it is used to train and evaluate the model for each of the Big Five dimensions.

In this paper, we argue that the subtle differences in language use and behavioral manifestations of each dimension might require a unique modelling approach. We perform extensive analysis to examine how a selection of linguistic markers, a classifier and a performance metric reflects on each Big Five dimension. Our objective was to deepen our understanding on the relationship between these choices and the model output from the viewpoint of various application scenarios. The analysis has highlighted that while some classifiers are more effective, robust and perform more consistently across conditions, there is no universally best classifier for all dimensions, and the initial choice of features does affect the predictiveness of each Big Five dimension.

In what follows, we survey the related research and present the models that may facilitate the personality prediction of vloggers. In the wake of the results from our experimentation, we present the general conclusions of our analysis and highlight the implications for future research.

2   Related Research

In a modern era of pervasive social networking, relying on traditional psychometric solicitation methods for assessing personality is not an idea to be dwelled

upon. It is rather the potential of intelligent technologies, such as, machine learning and data mining that are seen as a viable alternative to capture the observable manifestation of behavior trends and patterns pertaining to personality.

Establishing the theoretically- and empirically-supported conjecture that there is a strong correlation between the personality traits of a speaker or author of a written text and their language use has been of crucial relevance for the field. The Medical Research Council (MRC) Psycholinguistic Database [7], released in 1981, is one of the pioneering efforts. It contains semantic, syntactic, phonological and orthographic information on 98,538 words in the database, facilitating mapping between each personality trait and the tangible linguistic indicators. A little more than a decade later, a very instrumental text analysis tool, the Linguistic Inquiry and Word Count (LIWC) that includes a word dictionary, had emerged from the work of James W. Pennebaker and his group, [24]. The LIWC dictionary is comprised of words categorized in five main categories and more than 80 subcategories. Features derived from these lexicons are the cornerstone of many influential works on this topic.

The scarcity of large datasets labeled for ground-truth presents a bottleneck in studying language indicators for personality prediction. Crowd-sourcing and social networks have emerged as platforms to meliorate the problem. James W. Pennebaker and Laura A. King had compiled what has since become a widely-exploited essay dataset, composed of 2400 essays written by psychology students on subjects of their own choice and labeled manually by judges with scores for each of the Big Five personality traits [25]. Two datasets containing multimodal data exist, namely, the EAR dataset [16], a smaller corpus of conversation extracts recorded using an electronically activated recorder, and the IDIAP YouTube dataset, used in our study, consisting of 404 YouTube videos, transcribed and labeled with personality impressions along the five personality dimensions. The myPersonality project [6] dataset including both textual data (e.g., status updates) and Facebook metadata is no longer maintained and available, although a number of prominent research studies based on the Facebook dataset have contributed to the field.

Syntactic linguistic features such as part-of-speech tags (POS) and dependency parsing [22], lexicons [25], language modeling [23], psycho-linguistic [7,24] and their combinations [13] have been identified as tangible verbal indicators of personality traits. The performance of the models varies across datasets, with averaged accuracy on all Big Five dimensions ranging from 56% to 64% for the essays dataset, and between 57% and 73% on the EAR corpus [13]. Stronger predictive performance has been achieved on the myPersonality dataset, with precision scores between 86% and 95% across all Big Five dimensions [15]. A number of machine learning (ML) algorithms, such as SVM [13,20], logistic regression, discriminant analysis, decision trees [13] and deep learning [14], have been utilized in the personality predictive models.

A review of previous studies using the IDIAP YouTube dataset was carried out, in particular the papers presented at the Workshop of Computational Personality Recognition 2014 [5]. All models discussed in the six papers have

outperformed the baseline results - an F1 score of 39%, calculated as an average of the five F1 scores across the Big Five dimensions. The best-performing model using an affective analysis on a coarse- and fine-grain level has yielded an average F1 score of 72.7% [9].

3   Dataset

Our models were constructed on the IDIAP YouTube[1] personality dataset, which consist of 404 YouTube video monologues, vlogs, showing individuals (vloggers) of both gender, talking on a variety of topics in front of the camera. The behavioral features are comprised of 21 audio features, which include speech activity measurements and prosody cues (e.g., average and standard deviation of the voice pitch, voice rate, speaking time), and 4 video features describing the motion of the vlogger in the video (e.g., entropy, median and center of gravity in horizontal and vertical dimensions) [3]. Every video has been transcribed and labeled manually with gender and personality impressions for the vlogger along the Big Five personality dimensions solicited by crowdsourcing [4].

4   Methodology

We have developed several models, each of which consists of different combinations of diverse types of linguistic features. A limited preprocessing has been performed: removal of all special characters except apostrophes in order to preserve word contractions, conversion to lowercase letters and in some cases, removal of the stop words, if the initial testing had indicated their significant impact on the predictive performance.

4.1   Features

The types of features that were incorporated in our models have been previously utilized in research with objective related to ours, such as: affective analysis, detecting deception in text, modeling personality, although their selection and implementations as well as their place in the modeling pipeline differ. The predictive power of the linguistic features vary across tasks and contexts, and their selection follows a careful analysis and consideration for the specific requirements of the task at hand. The unifying aspect highlighted across related research is that the emotional content and language nuances, rather than understanding of the semantic meaning of the content are more important when modeling phenomena related to human nature. The 499 linguistic features extracted from the vlog transcripts and incorporated in our feature collections could be naturally categorized into 4 different types and a short description of each category follows.

---

[1] https://www.idiap.ch/dataset/youtube-personality.

Lexicons of Topic and Affective Words. The findings that extend across several studies is that positive, negative sentiment words and topical words, correlate highly with the Big Five personality traits [10,13,20,25]. The following six lexicons were used to capture the affective content and the category of topics present in transcribed vlogs:

Bing Liu's Opinion Lexicon[2]. A lexicon of almost 6,800 positive and negative words [12], generated by a dictionary- and corpus-based approach has been frequently used for sentiment analysis and opinion mining. In our models, for each transcript, stop words were removed and two features based on this lexicon were calculated, the frequency of positive and negative lexicon words found in the transcript.

SentiWordNet[3]. A lexical resource created for sentiment classification and opinion mining [2], by annotating each synset in WordNet [17] with three scores denoting the strength of positive, negative and neutral emotion carried by the terms in a synset. For each SentiWordNet term found in a vlog transcript, four features were calculated, namely the cumulative and maximal positive and negative score.

NRC Affect Intensity Lexicon[4]. A list of English words associated with their intensity scores for four basic emotions: anger, fear, sadness, joy [19]. Four features were calculated, one for each emotional category in the lexicon, namely, the frequencies of all NRC lexicon words detected in a transcript that has intensity score for each of the four emotions higher than 0.35.

NRC Valence, Arousal, and Dominance (VAD) Lexicon[5]. A lexicon of more than 20,000 English words annotated with their scores along three sentiment dimensions: valence, arousal, and dominance [18]. Three features, one for each of the VAD metrics, valence, arousal or dominance, were calculated as the frequencies of NRC VAD lexicon words that have a score higher than 0.5 for the corresponding metric.

Empath[6]. A tool that facilitates generation of personalized lexical categories based on a few seed terms provided by a user using deep learning architectures for mining the web and crowdsourcing [8]. Empath's categories have been found to correlate highly ($r = 0.906$) with humans i.e., similar categories in LIWC. Two hundred built-in categories on a number of common topics extracted from modern fiction are also available. For the purpose of this research, the frequency counts of all Empath built-in word categories identified in a transcript were accounted as features in our models.

---

[2] https://www.cs.uic.edu/~liub/FBS/sentiment-analysis.html#lexicon.
[3] https://github.com/aesuli/SentiWordNet.
[4] https://saifmohammad.com/WebPages/AffectIntensity.htm.
[5] https://saifmohammad.com/WebPages/nrc-vad.html.
[6] https://github.com/Ejhfast/empath-client.

Harvard Inquirer H4Lvd[7]. A dictionary of words that are assigned with 182 properties (e.g., emotion, strength, arousal, pain, etc.) [1] was originally developed for social-science content-analysis research. In our work, we have constructed features for each category by calculating its word frequency counts.

LIWC. Evidence has shown that a selection of psycholinguistic features can be useful as footprints of someone's style, emotional state and personality. What a lot of research in personality prediction have in common is the use of the tool LIWC[8], that includes a lexicon of words, divided into 5 categories and various number of subcategories. For example, linguistic category: nouns, pronouns, assents, negations; psychological process: anger, anxiety; relativity: past, present and future tense verbs, words depicting space and time; personal: achievement, religion, school, sports; and experimental dimension: swear words, nonfluencies and fillers. For our particular study, we have used the frequencies of 68 word categories. The average word count per sentence and the number of sentences in each transcript were also identified as important sentence-level psycholinguistic indicators of vlogger's talkativeness.

Language Modeling. Even though, research argues that the generalizability of the model might suffer, forty most frequent bigrams have been extracted, and their tf-idf values have been calculated for each vlog transcript.

At the onset of our explorations, we have investigated the correlation between the selected 499 lexical features and the class labels corresponding to the Big Five dimensions, and highlight the most significantly correlated ones. While lexicon words showed significant correlation with all personality traits, the effects of other types of features varied across traits. As would be expected, the correlation analysis has shown that every personality trait except Extraversion is correlated to features depicting negative emotions (e.g., sadness, anger) and use of swear words. Extraversion is associated with being outgoing, friendly and socially active, so the correlation with words belonging to related categories (e.g., journalism, writing, play, pleasure) is substantiated by the correlation results. Extraverts are considered to be talkative, hence the correlation to the average uttered words per sentence was found significant.

4.2 Feature Collections

As a baseline model we have adopted the initial study performed on the IDIAP dataset [3,4], which is based on the audio-visual features extracted and provided by the IDIAP dataset and the gender of the vlogger. No linguistic features were included in the baseline model. In order to analyze the relevance and predictive power of the four categories of linguistic features described in the previous section and their combinations, a large number of feature collections were devised and evaluated, but we limit our discussion to the following feature collections:

---

[7] http://www.wjh.harvard.edu/~inquirer/spreadsheet_guide.htm.
[8] http://liwc.wpengine.com/.

* [**A.**] Baseline feature collection.
* [**L1 - L7.**] Each of the L1 - L6 feature collections adds one of the lexicons we considered to the baseline model, with L7 adding all lexicons to the baseline feature collection.
* [**P1 - P3.**] The feature collections P1 to P3 use LIWC, sentence statistics, and both types of psycholinguistic features, respectively, in addition to the baseline features.
* [**B1.**] This feature collection complements the baseline model with the bigrams.
* [**F1 - F3.**] Selection of the 75 most relevant linguistic features according to the calculated information gain and chi-squared test were included in F1 and F2, respectively. A compilation of all 499 features added to the baseline were included in F3.

5   Discussion of Results

Six different classifiers have been tested: Multinomial Naive Bayes (MNB), Extra Trees (ET), Random Forest (RF), Support Vector Machines (SVM), Multilayer Perceptron (MP) and K-Nearest Neighbors (KNN). The classifiers have been trained and tested separately for each Big Five dimension, their performance was measured with the following metrics: accuracy (A), precision (P), recall (R), F1 score (F1), negative predictive value (NP) and specificity (S). The first four performance measures are the most popular and widely-used. The inclusion of two additional metrics was deemed necessary to support our discussions as both the predictiveness of positive and negative classes i.e. personality traits on opposite sides of a personality dimension are of interest. In particular, negative predictive value shows how many of the elements that are predicted as negative are actually negative, while specificity indicates how many of the actually negative elements are predicted as negative. The performance of the best-performing classifier-features combinations for each Big Five dimension are summarized in Table 1.

While all best-performing models consisting of a particular feature collection and a choice of machine learning algorithm outperform the baseline model in sensitivity (recall), specificity and F1 measure, the performance gains differ depending on the evaluation metric we observe as evident in Table 1. It was noted that the difference between the best and worst F1 value obtained for different personality dimension varies and can be as high as 28%. These variational patterns can be considered as important empirical evidence of the confrontation between the suitability of the model to capture the behavioral manifestations of different personality traits reflected in the language use. It appears that the inclusion of the Harvard Inquirer H4Lvd and NRC Affect Intensity Lexicon led to substantial performance gains for predicting Agreeableness and Neuroticism, respectively, compared to the baseline results, while the best models for the other three dimensions perform comparably. The performance differences for Extroversion and Neuroticism were less notable when compared to the baseline results

Table 1. Results for the best combinations of a classifier and features (including the baseline) are presented for each personality dimension. Extroversion - EXT, Agreeableness - AGR, Conscientiousness - CON, Neuroticism - NEU, Openness to experience - OPE.

	Features	Classifier	A	P	R	F1	NP	S
EXT	Gender + audiovisual (baseline)	ET	76.8	61.1	64.7	62.8	84.2	82.1
EXT	Baseline + LIWC	SVM	73.2	53.6	88.2	66.7	92.9	66.7
AGR	Gender + audiovisual (baseline)	SVM	67.9	65.6	75.0	70.0	70.8	60.7
AGR	Baseline + Harvard Inquirer H4Lvd	MNB	80.4	77.4	85.7	81.3	84.0	75.0
CON	Gender + audiovisual (baseline)	ET	76.8	76.4	100.0	86.6	100.0	7.1
CON	Baseline + NRC Affect Intensity lexicon	MNB	80.4	81.6	95.2	87.9	71.4	35.7
NEU	Gender + audiovisual (baseline)	SVM	58.9	50.0	60.9	54.9	67.9	57.6
NEU	Baseline + NRC Affect Intensity lexicon	SVM	67.9	58.6	73.9	65.4	77.8	63.6
OPE	Gender + audiovisual (baseline)	SVM	71.4	55.0	61.1	57.9	80.6	76.3
OPE	Baseline + all lexicons	SVM	73.2	57.9	61.1	59.5	81.1	78.9

than was the case for the other Big Five dimensions. The best performance for Extroversion was achieved by augmenting the baseline features with the categories from the LIWC, when using the SVM classifier, although the superiority against the baseline was not evident in every evaluation metric.

To get a better insight on how the performance varies with the selection of the word lexicon, we have compared the prediction performance across all feature collections and all Big Five dimensions. It is worth noting that the use of one lexicon, rather than another, yielded very different results, sometimes with a 15.3% difference in F1 score. The evaluation of the models that include the features from all lexicons, feature collection F3, was the best-performing model in combination with SVM only for the Openness to experiences dimension. It did not point to any performance advantage, although often it is expected that adding new features yields performance gains, rather than making it worse. We may speculate that the overlapping of word categories among lexicons unfavorably affected the performance and a compiled curated lexicon might be a solution.

The radar plots presented in Fig. 1 show the top 5 best-performing models for each personality dimension and support our previous analysis that the performance advantage contributed to a particular lexicon vary widely. The impact of the combination (feature collection, classifier) integrated in a model produces performance improvements or substantial reductions depending on the evaluation metric and the personality dimension we focus on, in some cases even by 21.7% (the recall for Neuroticism).

Possibly the most direct experimental evidence against the practice of choosing the same feature set and same classifier for modeling all personality dimensions comes from the following analysis that take different views and look at the impact a selection of classifier has on a particular performance metrics. A selection of the optimal classifier for all personality dimensions, leads to an unbalanced

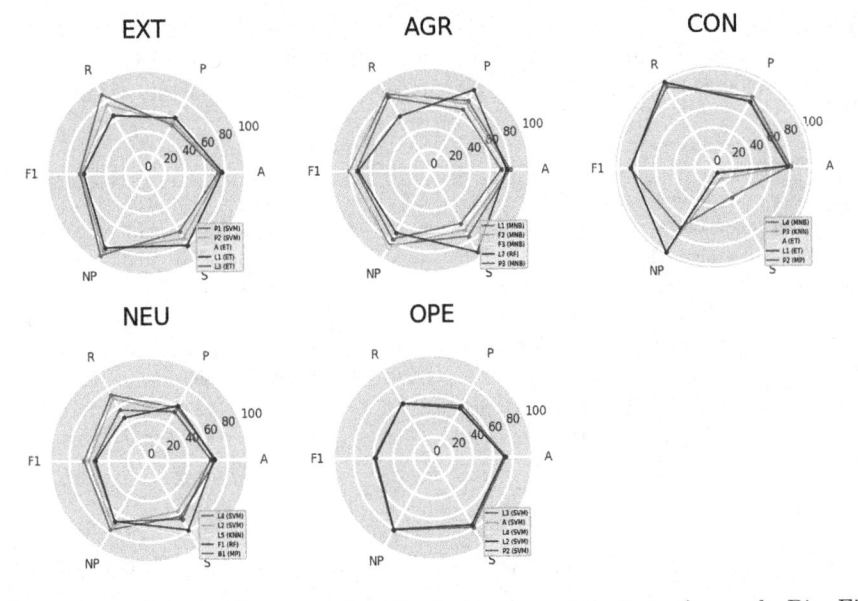

Fig. 1. Top 5 best-performing classifier-features combinations for each Big Five dimension.

performance, shaped by the differences in the underlying feature collection employed in each model as shown in the radar charts in Fig. 2. In general, these rather substantial variations means that a classifier can make better and more accurate predictions for one dimension, but fails another. Substituting one classifier with another (e.g., Multilayer Perceptron with SVM in our analysis) has promoted completely new sets of 5 top performing models and revealed different patterns across dimensions. For example, the best models for Openness to experience were L3, A, L4, L2 and P2 and the rather substantial differences evident in Fig. 2 diminished with the use of SVM. The patterns for the other dimensions were consistent, but the differences were less pronounced.

Our experiments make use of various types of features and their combinations, so in order to address the question of whether or not, and to what extent, a particular classifier would affect the performance results, we have selected the feature collection F3 that includes all 499 features of diverse nature considered in our study (Fig. 3). It was found that the choice of classifier makes a significant difference in the outcomes. For example, the choice of Multinomial Naive Bayes and feature collection F3 has achieved the best recall of 85.7% for the Agreeableness, compared to only 51.7% obtained by the SVM classifier. On the other hand, there was a dramatic decrease of 36.9% in the specificity between Multinomial Naive Bayes and Extra Trees when predicting Openness to experiences, utilizing the same feature collections F3.

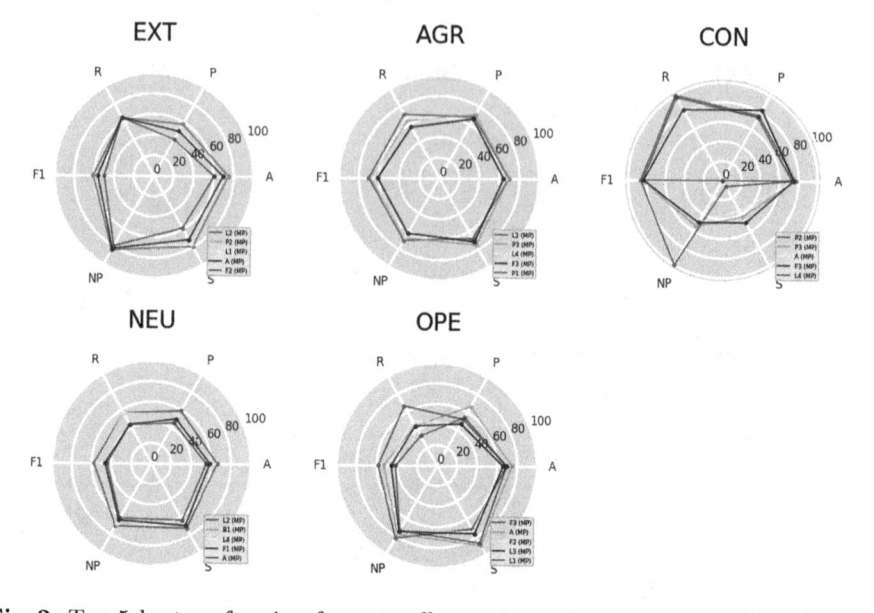

Fig. 2. Top 5 best-performing feature collections combined with the Multilayer Perceptron classifier for each Big Five dimension.

A quite different picture on classifier's performance arises from the analysis of consistency of their performance. An interesting trend was observed when looking at the averaged as opposed to peak performance of the classifiers, shown in Figs. 4 and 5. Multinomial Naive Bayes, SVM and Multilayer Perceptron have showed balanced performance across all measures for every Big Five dimension, except Conscientiousness, where all suffered at either negative predictive value or specificity. K-Nearest Neighbors, while also balanced across all performance measures, had only obtained one best result across all of the evaluation metrics and dimensions, and averaged scores lying on the bottom end of the spectrum.

We argue that it would also be valuable to track the roles certain choices played in improving the performance on one metric vs another and suggests that the established practice of choosing one feature collection for all personality dimension might have much higher impact than what is suggested by the empirical studies. A selection of evaluation metric should be driven by a particular scenario or application objective. Assessing recall (true positives) might be more suitable for identifying neurotic students in particular intelligent tutor that adapts interaction dialogue to student's personality, however avoiding false-labeling people might ask for tracking measures, such as F1, specificity and negative predictive value. A substantially different relationship might be of interest in a tool supporting and mediating collaboration of group of people with "clashing" personalities, as opposed to matching web design properties that appeal to certain personalities.

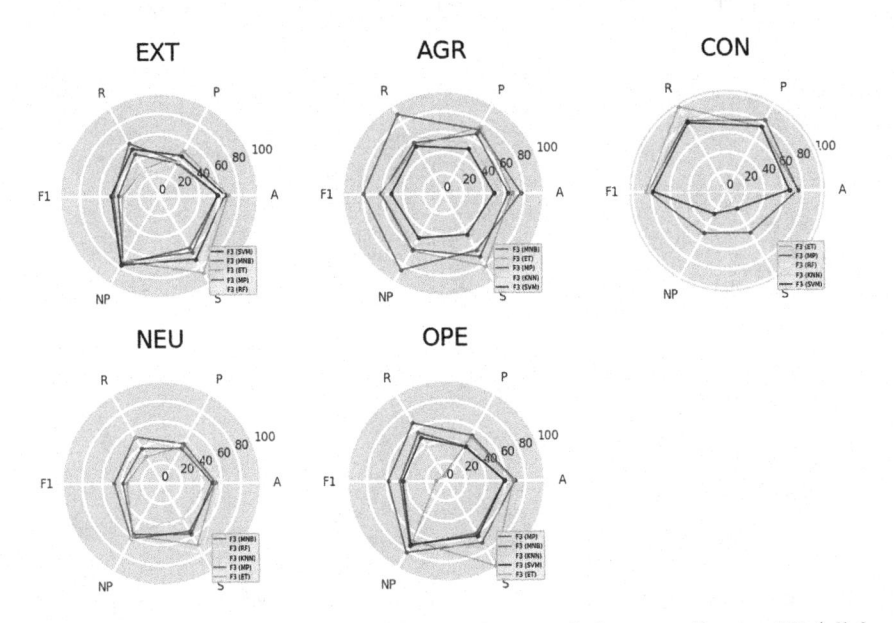

Fig. 3. Top 5 best-performing classifiers combined with feature collection F3 (all features) for each Big Five dimension.

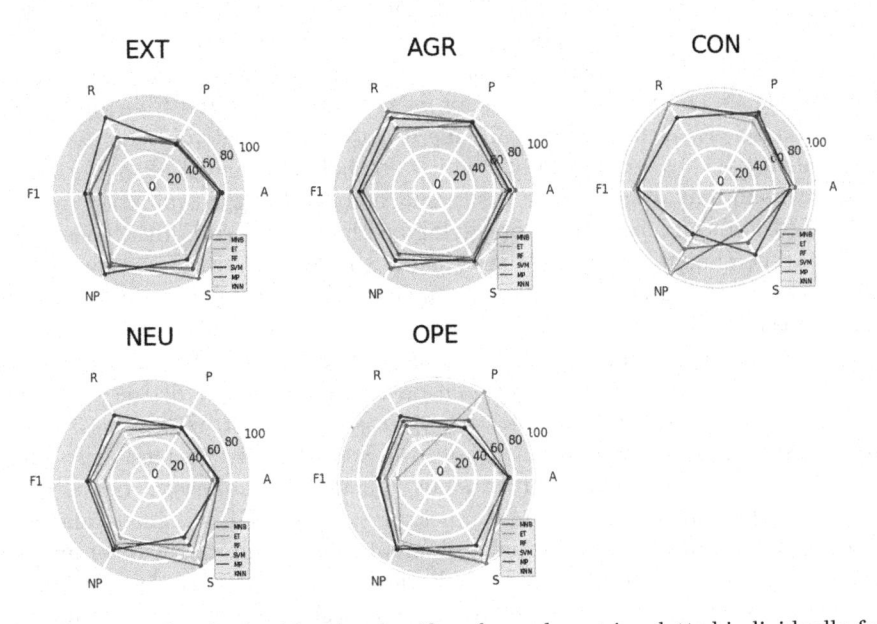

Fig. 4. Best results obtained by the classifiers for each metric, plotted individually for each Big Five dimension.

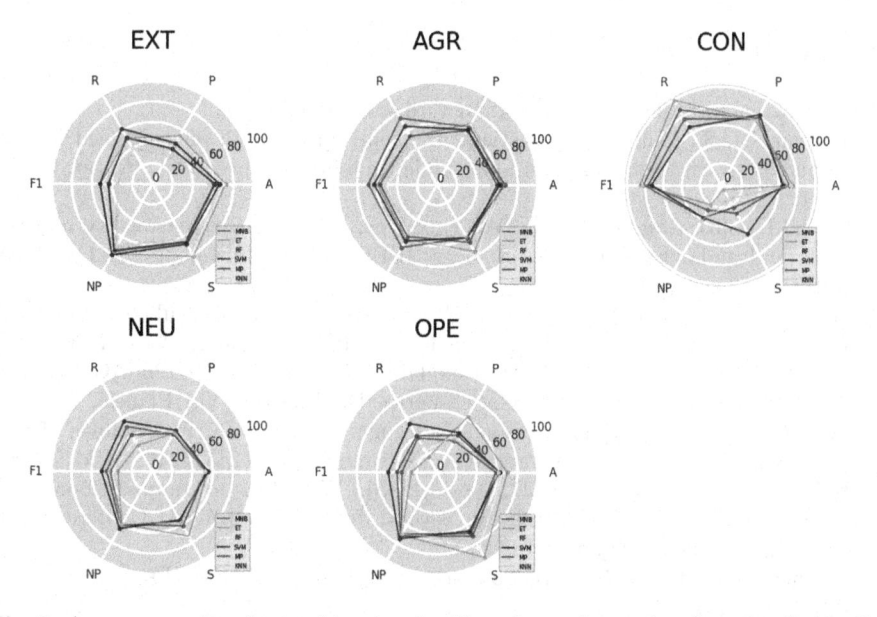

Fig. 5. Average results obtained by the classifiers for each metric, plotted individually for each Big Five dimension.

6   Conclusion

This paper reports on a study that investigates that interplay and trade-offs between the choice of classifier, feature collection and their combinations for the task of predictive modeling of personality in video monologues. Our analysis casts doubts on researchers' practice of choosing the same pair of classifier and feature collection for all Big Five dimensions that might potentially lead to miss-classifications of personality traits. The study also points to the importance of unavoidable trade-off between the choice of performance metric and the application scenario under investigation.

The subtle differences in behavioral manifestations of personality expressed in language use might require a unique modeling approach for each personality dimension. No single study can resolve all issues and offer solutions. We continue with the way we conduct scientific inquiries - extensive experimentation and in-depth analysis that might shed light on the causation rather than correlations found in data.

Acknowledgements. This work was partially financed by the Faculty of Computer Science and Engineering at the "Ss. Cyril and Methodius" University.

References

1. Harvard inquirer. http://www.wjh.harvard.edu/~inquirer/
2. Baccianella, S., Esuli, A., Sebastiani, F.: Sentiwordnet 3.0: an enhanced lexical resource for sentiment analysis and opinion mining. In: Lrec, vol. 10, pp. 2200–2204 (2010)
3. Biel, J.I., Gatica-Perez, D.: The YouTube lens: crowdsourced personality impressions and audiovisual analysis of vlogs. IEEE Trans. Multimedia **15**(1), 41–55 (2013)
4. Biel, J.I., Tsiminaki, V., Dines, J., Gatica-Perez, D.: Hi YouTube!: personality impressions and verbal content in social video. In: Proceedings of the 15th ACM on International Conference on Multimodal Interaction, pp. 119–126. ACM (2013)
5. Celli, F., Lepri, B., Biel, J.I., Gatica-Perez, D., Riccardi, G., Pianesi, F.: The workshop on computational personality recognition 2014. In: Proceedings of the 22nd ACM International Conference on Multimedia, pp. 1245–1246. ACM (2014)
6. Celli, F., Pianesi, F., Stillwell, D., Kosinski, M.: Workshop on computational personality recognition: shared task. In: Seventh International AAAI Conference on Weblogs and Social Media (2013)
7. Coltheart, M.: The MRC psycholinguistic database. Q. J. Exp. Psychol. Sect. A **33**(4), 497–505 (1981)
8. Fast, E., Chen, B., Bernstein, M.S.: Empath: understanding topic signals in large-scale text. In: Proceedings of the 2016 CHI Conference on Human Factors in Computing Systems, pp. 4647–4657. ACM (2016)
9. Gievska, S., Koroveshovski, K.: The impact of affective verbal content on predicting personality impressions in YouTube videos. In: Proceedings of the 2014 ACM Multi Media on Workshop on Computational Personality Recognition, pp. 19–22. ACM (2014)
10. Gievska, S., Koroveshovski, K., Chavdarova, T.: A hybrid approach for emotion detection in support of affective interaction. In: 2014 IEEE International Conference on Data Mining Workshop, pp. 352–359. IEEE (2014)
11. John, O.P., Srivastava, S., et al.: The Big Five trait taxonomy: history, measurement, and theoretical perspectives. Handb. Pers. Theor. Res. **2**(1999), 102–138 (1999)
12. Liu, B., Hu, M., Cheng, J.: Opinion observer: analyzing and comparing opinions on the web. In: Proceedings of the 14th International Conference on World Wide Web, pp. 342–351. ACM (2005)
13. Mairesse, F., Walker, M.A., Mehl, M.R., Moore, R.K.: Using linguistic cues for the automatic recognition of personality in conversation and text. J. Artif. Intell. Res. **30**, 457–500 (2007)
14. Majumder, N., Poria, S., Gelbukh, A., Cambria, E.: Deep learning-based document modeling for personality detection from text. IEEE Intell. Syst. **32**(2), 74–79 (2017)
15. Markovikj, D., Gievska, S., Kosinski, M., Stillwell, D.J.: Mining facebook data for predictive personality modeling. In: Seventh International AAAI Conference on Weblogs and Social Media (2013)
16. Mehl, M.R., Gosling, S.D., Pennebaker, J.W.: Personality in its natural habitat: manifestations and implicit folk theories of personality in daily life. J. Pers. Soc. Psychol. **90**(5), 862 (2006)
17. Miller, G.A.: WordNet: a lexical database for English. Commun. ACM **38**(11), 39–41 (1995)

18. Mohammad, S.: Obtaining reliable human ratings of valence, arousal, and dominance for 20,000 English words. In: Proceedings of the 56th Annual Meeting of the Association for Computational Linguistics (Volume 1: Long Papers), vol. 1, pp. 174–184 (2018)
19. Mohammad, S.M.: Word affect intensities. arXiv preprint arXiv:1704.08798 (2017)
20. Mohammad, S.M., Kiritchenko, S.: Using nuances of emotion to identify personality. In: Proceedings of ICWSM (2013)
21. Myers, I.B., McCaulley, M.H., Quenk, N.L., Hammer, A.L.: MBTI Manual: A Guide to the Development and Use of the Myers-Briggs Type Indicator, vol. 3. Consulting Psychologists Press, Palo Alto (1998)
22. Oberlander, J., Gill, A.J.: Individual differences and implicit language: personality, parts-of-speech and pervasiveness. In: Proceedings of the Annual Meeting of the Cognitive Science Society, vol. 26 (2004)
23. Oberlander, J., Gill, A.J.: Language with character: a stratified corpus comparison of individual differences in e-mail communication. Discourse Process. **42**(3), 239–270 (2006)
24. Pennebaker, J.W., Francis, M.E., Booth, R.J.: Linguistic inquiry and word count: LIWC 2001. Mahway: Lawrence Erlbaum Associates **71**(2001), 2001 (2001)
25. Pennebaker, J.W., King, L.A.: Linguistic styles: language use as an individual difference. J. Pers. Soc. Psychol. **77**(6), 1296 (1999)

Ski Injury Predictions with Explanations

Sandro Radovanović$^{(\boxtimes)}$ ⓘ, Andrija Petrović ⓘ, Boris Delibašić ⓘ,
and Milija Suknović

Faculty of Organizational Sciences, University of Belgrade, Jove Ilića 154,
Belgrade, Serbia
sandro.radovanovic@fon.bg.ac.rs

Abstract. Providing prediction models for ski injuries is a very challenging classification problem. We propose a model for injury prediction that uses ski lift trajectory features. Ski slopes, in general, differ by width, length, difficulty and geographical position on the mountain, which results in different patterns of skiing. We study the correlation between these patterns different types of ski injuries. Many types of analysis were proposed in this domain of research. However, they are either too simple for real-time usage, such as univariate statistical analysis, or use interpretable predictive models at the cost of lowering accuracy. In order to gain best predictive performance and still provide explanation one must combine different approaches. We utilize modern algorithms such as random forests and gradient boosted trees with explainability methods Shap and Lime for providing interpretation about reasons for specific decision. The proposed models were created on Mt. Kopaonik, Serbia ski resort and it is shown that ski injury in the following hour on specific ski slope can be predicted with AUC ~ 0.76, which is better up to $\sim 15\%$ compared to classical approaches such as logistic regression and decision trees.

Keywords: Ski injury prediction · Machine learning · Explainability · Shapley value · Lime

1 Introduction

From all entertainment industries worldwide ski sports and leisure industry is among the leaders with approximately 400 million of skier visits worldwide [14]. This number is steady over the years, which indicates that ski resorts are developing and fulfilling skiers needs. Having that amount of skier visits, leads to unpleasant situations, such as injuries. Although injuries in ski resorts occur seldom, the costs and outcomes of such injuries can be high and significant. For example, injury often leads to temporary or permanent work disability, movement disability or even death [2]. Because of that, it is of critical importance that risk factors are monitored constantly and that injury is identified in short time span. Usually, ski resorts have implement safety strategies, but unwanted events still occur. Namely, decisions where injury is going to occur and who is going to be injured are made based on experience, intuition or opinion. This leads to non-optimal solutions which can be improved by utilizing existing data.

Every modern ski resort has RFID ski passes for entering ski lift gates. Having this technology one can obtain behavior data for each skier, i.e. how many slopes a day a

© Springer Nature Switzerland AG 2019
S. Gievska and G. Madjarov (Eds.): ICT Innovations 2019, CCIS 1110, pp. 148–160, 2019.
https://doi.org/10.1007/978-3-030-33110-8_13

skier does, how fast is he or she skiing or whether skiers tend to often change ski slopes. All these derived behavior attributes can be combined with existing injury data with the goal of developing predictive models. This allows usage of modern data analysis methods such as predictive modelling and machine learning algorithms. Unfortunately, ski injury research is often done on small-scale, case-control studies that are used for analysis of injured population compared to small sample of non-injured skiers. Additionally, self-reported information about skiing experience, skiing performance are commonly used. Analysis is done using descriptive statistics or statistical hypothesis testing.

The problem with ski injury prediction is that ski injuries are very rare events with 0.2% or fewer injuries per skier day [22]. Although the percentage is very low number of skiers that are injured is around 800,000 on worldwide level. This poses a major public health problem, which lowers quality of life of an individual and presents a major cost for insurance companies.

Predictive modeling and machine learning have been applied in ski injury prediction already. Majority of these papers dealt with the task of making correct predictions or explaining what subpopulation is more prone to injury. However, to the best of our knowledge none of the papers dealt with explanation of why an individual is prone to injury. Therefore, in this paper we propose a predictive model for identification of ski injury on slope and for specific time interval, for specific individuals. Additionally, we present several methods for explaining the decision about whether an injury is going to occur. This way we investigate risk factors driving injury related behavior which allows decision maker to use results are decision support system.

In this paper we created and evaluated several traditional and state of the art predictive models for binary classification task of predicting whether an injury will occur in the following hour on specific ski lift on Mt. Kopaonik, Serbia. This allows identification of risk factors and real-time response and prevention of injuries. As an evaluation measure, we used a predictive performance measure that is commonly used for this task, called area under the ROC curve (AUC).

The remainder of the paper is structured as follows. Section 2 provides a literature review on ski injury predictions. Section 3 provides a methodology of the research providing a brief description of the data, experimental setup, evaluation measures and explainability techniques. Section 4 present results and discussion of the result, while Sect. 5 concludes the paper.

2   Literature Review

Analysis of ski injury data is often done using descriptive statistics, basic statistical hypothesis testing or using correlation between self-reported questionnaire data. This is referred in literature as small-scale, case-control studies. The idea of such studies is to compare sample of injured to sample of non-injured skiers and inspect whether there are differences between different characteristics, properties or behavior of skiers. As a result, one can find average values, standard deviations, correlations and most often

odds ratios or risk ratios. Common analysis is done according to physical properties of a skier such as gender [20] and age [6, 19]. Also, skiing behavior is inspected, i.e. speed of skiing [20] or skiing experience [12]. Finally, one can find analysis of ski resort properties, quality of snow [21], weather [12] and ski lift quality [6].

One can discuss usability of this analysis. Namely, findings from above-mentioned studies can be used for educational purposes, i.e. for teaching future skiers about risk factors or as information for rescue service but real-time decision-making is very limited. In more up to date papers computational methods such as data mining and machine learning methods are used. In paper [4] multi-criteria decision-making method DEX was used combined with decision tree classifier method was used for prediction of global daily prediction of ski injuries. The mixture of domain knowledge and knowledge induced from data improved predictive performance and provided actionable insights for rescue service for capacity planning.

Decision tree classifier, more specifically, CHAID algorithm was used in paper [9]. The benefit of using decision tree algorithms compared to classical analysis in terms of odds ratio is that decision trees are able to find interactions of the attributes which leads to higher odds of injury. That identified interaction present subpopulation of skiers with specific properties (i.e. fast skiers that change slopes are more prone to injury compared to average skier). Additionally, CHAID algorithm has shown comparable results in terms of AUC compared to logistic regression algorithm.

Adding domain knowledge in order to provide better performance is presented in paper [8]. Domain knowledge is added in a stacking manner, more specifically predictive model is presented as a hierarchy of logistic regressions. Hierarchy is obtained from DEX models where each logistic regression can be seen as independent expert. Multiple predictions are combined using new logistic regressions up to final node of hierarchy. The framework can be seen as a feature extraction tool since each level in hierarchy provides different view on the problem.

However, ski injuries can be of different type as well. Therefore, decision support systems should also predict type of injuries as well. This can be regarded as a multi-label classification problem. In terms of data mining and machine learning one should utilized multi-label algorithms. In paper [15] one such application is presented. Another interesting approach can be found in [10] where instead of using classification models to predict whether an injury will occur or not one can use recommender systems. This way model recommends which ski slope is likely to have injuries. Although this seems less intuitive it has been shown that predictive performance of recommender systems was comparable or better than data mining and machine learning algorithms. Finally, one can find application of multi-task prediction models [16] where each ski slope is considered as a task.

More throughout analysis of ski injuries and predictive modelling can be found in [7, 17].

3   Methodology

In this section, we will present data, explanation and motivation of experimental setup.

3.1 Data

Data used in this research are from Mt. Kopaonik, Serbia ski resort. Mt. Kopaonik is the largest ski resort in Serbia with 20 ski lifts with different degrees of difficulty. In this paper we use data from the 14 most utilized ski lifts for which we could extract all features shown in Table 1. The dataset includes all ski lift gate entrances from season 2005/2006 to season 2011/2012. Ski lift entrances are stored using RFID check-ins of ski lift tickets at the beginning of the ski lift. Term skier refers to all people that use the ski lift transportation system, such as skiers, snowboarders etc.

Ski injuries are collected from Serbian mountain rescue service database while weather data is obtained from Republic hydro-meteorological service of Serbia. Final dataset is joined based on ski ticket numbers and date-time which resulted in a dataset of approximately 20 million rows. Since our goal is to make a prediction for an injury on ski slope for the following hour dataset is aggregated on ski slope, hour level which resulted in 44,941 observations. The dependent attribute (label) is converted to a binary value where value 1 means that at least one injury occurred in the following hour on ski slope, while value 0 means that a ski slope did not have any injury in the next hour. Aggregation of data allowed us to derive insightful attributes that describes typical skier behavior for that ski slope and also ski slope behavior.

Attributes can be roughly divided into four categories. First, we have descriptive statistics which is used to capture number of skiers on the ski slope. Second group of attributes present statistical properties of distribution such as average value, minimum, maximum, median, various percentiles, standard deviation, skewness and kurtosis for several distributions of data. Third group contains chemometric analysis of distributional data which is used to measure levels of activity in data [24]. This group of attributes contains the number of modes in distribution, the number of peaks and of pits, and the number of significant peaks. Finally, the weather data contains the temperature, dew point, humidity, visibility, fog, rain, snow, and thunder feature. There are in total 78 features.

Table 1. Feature explanation.

Feature group	Feature
Descriptive statistics	Total number of ski lift visited by skiers on ski slope
	Number of skiers on ski slope
Statistical	Average for feature group*
	Minimum value for feature group*
	Tenth percentile for feature group*
	First quartile for feature group*
	Median for feature group*
	Third quartile for feature group*
	Maximum value for feature group*
	Standard deviation for feature group*
	Skewness of distribution for feature group*
	Kurtosis of distribution for feature group*
	Number of modes in distribution for feature group*

(continued)

Table 1. (*continued*)

Feature group	Feature
Chemometric [24]	Number of turns in distribution for feature group*
	Signal that majority of the skiers are at the beginning of the distribution for feature group*
	Number of peaks in distribution for feature group*
	Number of pits in distribution for feature group*
	Number of significant peaks (p > 0.5) in distribution for feature group*
Weather	Temperature on ski slope in degrees Celsius
	Dew point on ski slope in degrees Celsius
	Humidity of air in %
	Visibility in kilometers
	Indicator for fog
	Indicator for rain
	Indicator for snow
	Indicator for thunder

*Statistical and Chemometric features are extracted for time spent on track, vertical distance of skier, number of ski slopes skier skied, and number of distinct ski slopes skier skied

3.2 Predictive Model Explainability

As explanation of predictions are a topic of rising interest in many application areas [18], which is also the case of ski injury predictions, we utilized Shapley values, and Lime. Shapley values explain prediction of an instance using game theory. Namely, each instance is a player in a game where predictions are considered as a payout. Then, by using coalitional game theory, one can calculate fair distribution of payout among the attributes for a player [5]. This explanation model belongs to the class of additive feature attribution methods which explain an output as a sum of attribution for each input attribute. More specifically, Shapley values tries to model equation presented in Eq. 1.

$$g(z') = \phi_0 + \sum_{i=1}^{M} \phi_i z_i' \tag{1}$$

where $z' \in \{0,1\}^M$, M is the number of input attributes, and ϕ_i presents attribute attribution value. Important property of such model should be local accuracy, missingness and consistency. Local accuracy means that attribute attributions are accurate in explaining predictions. Missingness explains that attributes that are not present in instance of interest generate zero value of importance and consistency present monotonicity of the output constraint. In other words, attribution method should be monotone function. In order to compute Shapley value we need to calculate attribute attribution values ϕ_i for each attribute. They are calculated using Eq. 2.

$$\phi_i = \sum_{S \subseteq N \setminus \{i\}} \frac{|S|!(M - |S| - 1)!}{M!} [f_x(S \cup \{i\} - f_x(S))] \tag{2}$$

where S is the set of non-zero indexes in z', and N the set of all input attributes. It is shown that this is only possible explanation method that satisfies all three properties (local accuracy, missingness and consistency) [13].

Having this kind of model we can obtain attribution of each attribute for prediction. Positive value of ϕ_i indicate that attribute i has positive impact on prediction. In this paper this will indicate that attribute i indicate that ski injury will occur in the following hour on specific ski slope for an instance we are looking for. Negative value of ϕ_i describes negative attribution of attribute i for prediction.

As an advantage of proposed method for explanation of predictions one can say that attribution of the attributes is fairly distributed and it is considered in areas that require explainability by law. Only Shapley value might be legally compliant method because of fairness property. Additionally, Shapley value allows contrastive explanation. This means that we can compare predictions to the average predictions on whole dataset level, data subset level or even on the instance level [1].

Although the method is mathematically sound and satisfies all desired properties practical use is limited because of exponential complexity of Eq. 2. However, for tree like models and ensembles of trees like models one can use algorithm described in [3].

Besides having individual attribute attribution, one can gain additional insight into problem at hand by adding interaction effects. Shapley values for decision tree like models can be obtained using Shapley interaction index [13] that is presented in Eqs. 3 and 4.

$$\phi_{i,j} = \sum_{S \subseteq N \setminus \{i,j\}} \frac{|S|!(M - |S| - 2)!}{2(M - 1)!} \nabla_{ij}(S) \tag{3}$$

where $i \neq j$, and

$$\nabla_{ij}(S) = f_x(S \cup \{i,j\}) - f_x(S \cup \{i\}) - f_x(S \cup \{j\}) + f_x(S) \tag{4}$$

For more detailed explanation authors refers to [21].

Other approach to model explainability is by using local surrogate models. This kind of models are used to explain predictions in general regardless of predictive model (whether it is complex or simple). In this paper we utilized local interpretable model-agnostic explanations (Lime). The idea of Lime is to train surrogate model which, instead of training general model for all instances, focus on training local surrogate models which are able to explain prediction of an instance. Mathematically, we can present Lime as in Eq. 5.

$$\exp(x) = argmin_{g \in G} L(f, g, \pi_x) + \Omega(g) \tag{5}$$

where x present an instance in dataset, L is loss function which is mean squared loss of prediction of an original model f and explanation model g with proximity measure π_x (how large neighborhood around instance x a model will look). Function $\Omega(g)$ measures model complexity. If this part is omitted explanation of a prediction for an instance will use many features which is an undesired property. However, using many attributes for an explanation provides better values for the loss function L. Therefore,

one must trade-off between accuracy and explainability. Finally, we might have unlimited number of explanations G and we must choose that function (model) g which is the best in terms of combination of loss L and complexity Ω [1, 11].

The process starts by selecting an instance for which we want an explanation. For that instance we perform bootstrap sampling several times and weight instances according to the Euclidean similarity to the instance of interest. Based on proximity measure we train a new model that is interpretable by nature, in this paper linear regression, using weighted instances. This way we obtain explanation of the predictions using interpretable models. It is worth to say that linear regression used lasso regularization in order to generate explanations which are not complex. In other words explanation is presented with a small number of attributes.

3.3 Experimental Setup

In order to provide best performing predictive model and provide explanation of predictions we utilized gradient boosted trees and random forest algorithms. Results are compared to logistic regression (lasso and ridge) and decision tree algorithms as they are commonly used in literature for this kind of problems. They provided better performance compared to traditional data mining and machine learning algorithms such as logistic regression and decision trees. However, they haven't been used due to lack of interpretability of results and explainability for one example. In order to provide realistic measure of performance we used cross-validation. More specifically we used 10 fold cross-validation, which means that the dataset was randomly split into 10 chunks, from which 9 chunks were used for model training and remaining one for model testing. The process was repeated 10 times, such that each chunk of data was used exactly once for testing. After finishing cross validation, 10 values of predictive performance were obtained and we present average value and standard deviation of predictive performance measure.

As a performance measure metrics we used AUC since it is a decision threshold independent measure (which means that AUC present the overall goodness of the predictive model) and because it is commonly used for evaluation of ski injury prediction models. AUC is calculated by calculating the sensitivity and specificity for every possible decision threshold available in the data. After creation of the curve we calculate the area under the curve. It can be calculated more easily using Mann-Whitney U test. AUC ranges from 0 to 1, where 1 presents a perfect classifier, while the value of 0.5 presents a classifier which is equal to the random classifier [23].

It is worth to notice that both random forest and gradient boosted trees have optimized parameters (number of trees) using internal 10 fold cross validation. Same applies for baseline methods, namely lasso and ridge logistic regression and decision tree.

Besides using predictive performance in terms of AUC we will provide interpretation and explanation of the predictions using Shapley values for instance, Shapley value for interactions and Lime.

4   Results and Discussion

Predictive performance is presented in Table 2. As described in Sect. 3 we trained and evaluated random forest and gradient boosted trees as complex algorithms which lack interpretability and explainability and compared predictive performance with baseline methods (algorithms commonly used for ski injury predictions) logistic regression with lasso and ridge regularization and decision tree algorithm.

Table 2. Predictive performance.

Algorithm	AUC (average ± standard deviation)
Lasso logistic regression	0.5257 ± 0.0971
Ridge logistic regression	0.6034 ± 0.0995
Decision tree	0.5136 ± 0.0336
Random forest	0.6885 ± 0.1108
Gradient boosted trees	0.7623 ± 0.1182

As we can observe gradient boosted trees obtained best predictive performance with AUC equal to 0.7623 which is better than random forest by ∼8% and baseline models for ∼15% up to ∼25%. This improvement in predictive performance is sometimes useless in real-life decision-making situations if the decision maker cannot understand how a predictive model makes decisions. Both random forest and gradient boosted trees are considered as algorithms that lack interpretability and explainability. We would like to use this improvement in predictive performance and still be able to provide explanation why ski slope is considered to be at risk of an injury. For that purposes, we utilize Shap and Lime methods explained in previous section.

Shap method provides us with explanation for an instance with individual contribution of each attribute. Example of Shapley values for two instances is presented in Fig. 1. Upper instance is example of ski slope in risk of having injury. The red arrows present attributes and their individual contribution to risk behavior for gradient boosted tree algorithm while blue arrows present lower risk for attribute attribution. Attribute attribution values for ski injury occurrence are sorted according the strength of attribution. The biggest attribution for that particular instance has skewness of distribution of vertical distance that skier skied up to a specific hour on a specific ski slope. The next best attribution is the number of unique (distinct) skiers on that ski slope. This is an indicator that ski slope is having a lot of skiers that just began skiing and small number of them which are skiing a longer period of time. Additionally, the ski slope is crowded with people. This kind of information is useful for injury mitigation. Other instance (lower part of Fig. 1) presents a ski slope for specific hour that is in no risk of having an injury in the following hour. As we can observe, some attributes indicate injury behavior (red arrows) while a lot of others indicate low risk behavior (blue arrows). The biggest injury risk attribute attribution is associated with kurtosis of number of slopes that skier had which means that skiers are not uniformly distributed in terms of number of slopes. However, we can also observe that we have a lot turns in

distribution and a lot peaks in distribution. This is probably the morning part of a skiing session where the ski slope is not crowded. Interpretation is in accordance with global finding in ski injury research [12] but this interpretation allows for more detailed inspection for each instance, or in this paper for each ski slope for each hour.

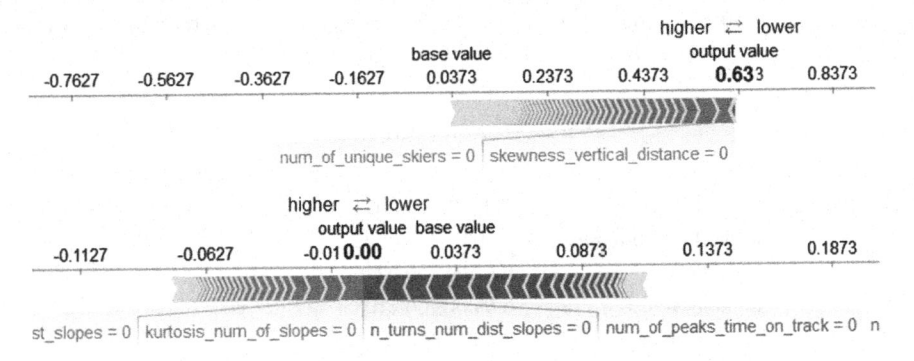

Fig. 1. Shapley values examples. (Color figure online)

Different point of view can be obtained from Lime. For the specific examples we can observe which attribute contributed most for decision. Blue value present non-injury behavior and orange color present injury related behavior. For this example, output for Lime model is presented on Fig. 2. It can be observed that average time on track influenced the most for this example. Its attribution is highest for injury behavior. However, overall behavior indicates that higher probability is that there will be no injury on this ski slope and on that hour.

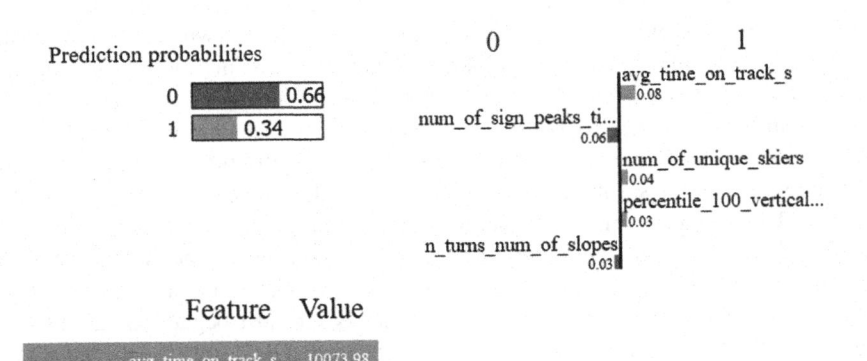

Fig. 2. Lime values examples. (Color figure online)

Shapley values can be inspected in different manner. Namely, one can inspect for each instance Shapley value for specific attribute and relate it to output value. This allows visual inspection for each attribute. In some sense it allows identification of areas where attribute influence predictions in non-linear manner.

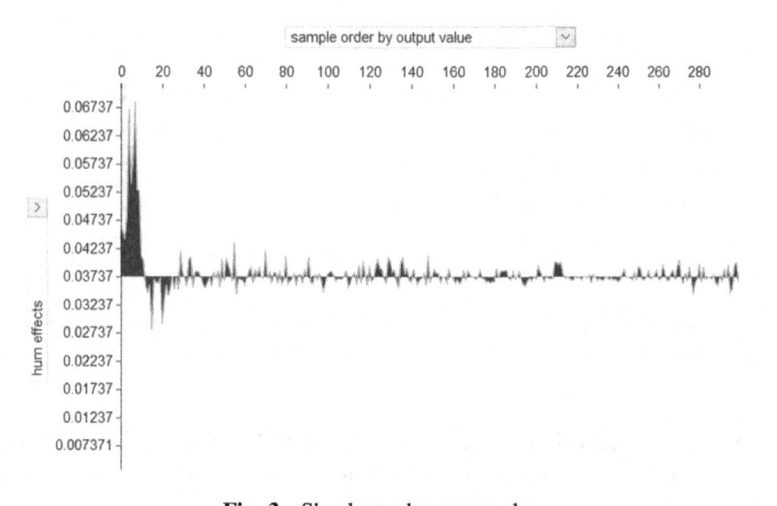

Fig. 3. Shapley values examples.

Another example is presented in Fig. 3. We inspected effects of humidity on ski injuries. On horizontal axis instances are sorted according to predictions in descending order, while on vertical axis attribution of humidity to predictions are presented. As we can observe for higher prediction of risk of injury humidity had high Shapley value which indicate that high humidity is a risk factor. However, as humidity gets lower it is unstable in Shapley value meaning that humidity after some point does not influence occurrence of injury. Similar applies to other weather condition attributes available in data, such as visibility, wind speed, temperature and dew point. To some extent this finding is in accordance with existing knowledge of influence of weather on ski injuries available in paper [12].

Finally, we can present Shapley interaction scores. For each feature we can create scatter plot and visually inspect obtained results. For the purposes of this paper, we present on Fig. 4 Shapley interaction scores for average time on track and number of distinct skiers presented. This combination is selected because average time on track has overall biggest interaction with number of distinct skiers. As we can observe average time on track (horizontal axis), number of unique skiers (color scale) seems not correlated, and Shapley interaction values are relatively low (vertical axis). However, at the beginning of the distribution we can observe that a higher number of distinct skiers are present mostly for low values of average time on track and we can also observe that Shapley interaction values are lower than zero which indicate non-injury behavior. Injury behavior occurs seldom for higher values of average time on track but it seems unrelated to the number of distinct skiers as well. This can be useful finding for this ski resort, because time spent on track is often a good predictor of ski injury occurrence.

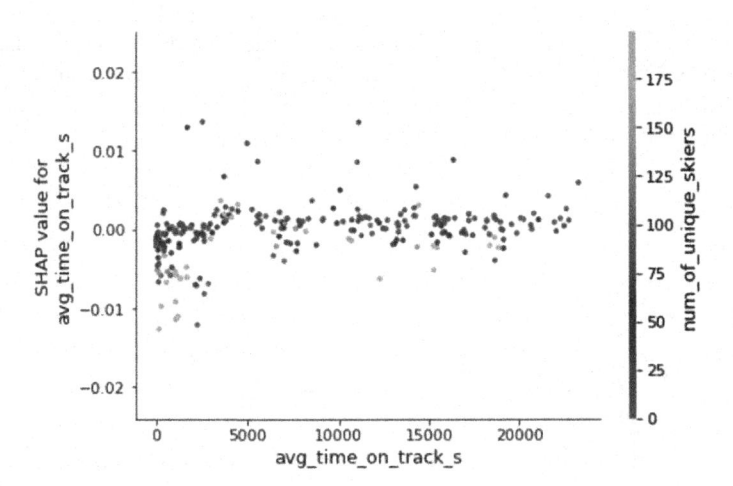

Fig. 4. Shapley interaction values for average time on track and number of distinct skiers. (Color figure online)

An interesting interaction is presented on Fig. 5. Shapley interaction values are presented for combination of temperature and number of pits (local minima) of vertical distance. Number of pits of vertical distance is selected because this attribute has best interaction score with temperature overall. We can observe that Shapley value increases as temperature increases (positive correlation). Number of pits vertical distance also has positive correlation and therefore and it seems that combination of these two attributes are jointly contributing to ski injuries.

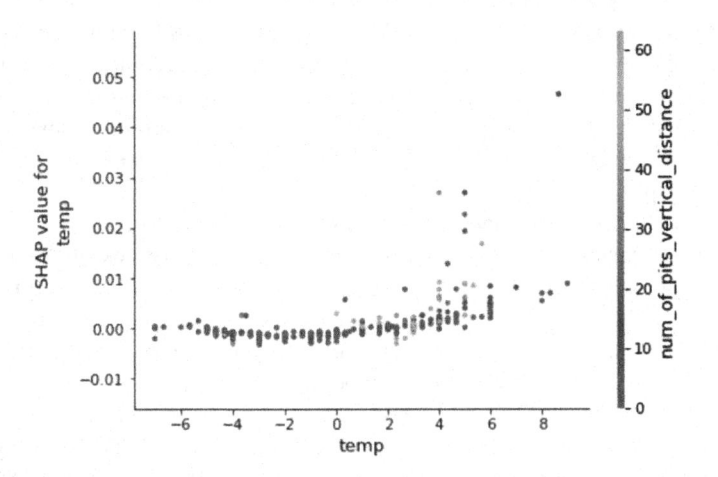

Fig. 5. Shapley interaction values for temperature and number of pits of vertical distance.

5   Conclusion

This paper proposes utilization of complex algorithms for ski injury prediction without hampering interpretability and explainability of predictive model. In this paper we trained and tested machine learning algorithms such as random forest and gradient boosted trees and compared them to logistic regression and decision tree algorithms. The problem we were dealing with was prediction of ski injuries on ski slopes for the one-hour time period. This problem is considered as a very hard binary classification problem mainly due to fact that ski injuries are rare events. Although they are rare events, cost of injury is very high. Because of that description of the prediction, or explanation of prediction, is necessary.

Predictive performance of gradient boosted trees is far better compared to algorithms used in existing literature such as logistic regression and decision trees. Namely, obtained AUC is 0.7623 and this number from ~15% up to ~25% better compared to baselines. The cost of utilization of more complex algorithm is alleviated using Shapley values and Lime method for explanation of predictions for each instance.

Based on Shapley values and explanations obtained from Lime model one can inspect influencing factors that lead to ski injury and provide necessary set of actions in order to prepare, act or prevent occurrence of ski injury. Besides inspecting just one attribute, decision maker can gain insight on influence of range of values of an attribute or interaction of attributes for ski injury occurrence.

References

1. Alvarez-Melis, D., Jaakkola, T.S.: On the robustness of interpretability methods. arXiv preprint arXiv:1806.08049 (2018)
2. Bianchi, G., Brügger, O., Niemann, S.: Skiing and snowboarding in Switzerland: trends in injury and fatality rates over time. In: Scher, I.S., Greenwald, R.M., Petrone, N. (eds.) Snow Sports Trauma and Safety, pp. 29–39. Springer, Cham (2017). https://doi.org/10.1007/978-3-319-52755-0_3
3. Biecek, P.: DALEX: explainers for complex predictive models in R. J. Mach. Learn. Res. 19(1), 3245–3249 (2018)
4. Bohanec, M., Delibašić, B.: Data-mining and expert models for predicting injury risk in ski resorts. In: Delibašić, B., et al. (eds.) ICDSST 2015. LNBIP, vol. 216, pp. 46–60. Springer, Cham (2015). https://doi.org/10.1007/978-3-319-18533-0_5
5. Casalicchio, G., Molnar, C., Bischl, B.: Visualizing the feature importance for black box models. In: Berlingerio, M., Bonchi, F., Gärtner, T., Hurley, N., Ifrim, G. (eds.) ECML PKDD 2018. LNCS (LNAI), vol. 11051, pp. 655–670. Springer, Cham (2019). https://doi.org/10.1007/978-3-030-10925-7_40
6. Chamarro, A., Fernández-Castro, J.: The perception of causes of accidents in mountain sports: a study based on the experiences of victims. Accid. Anal. Prev. 41(1), 197–201 (2009)
7. Delibašić, B., Radovanović, S., Jovanović, M.Z., Suknović, M.: Improving decision-making in ski resorts by analysing ski lift transportation—a review. In: Mladenović, N., Sifaleras, A., Kuzmanović, M. (eds.) Advances in Operational Research in the Balkans. SPBE, pp. 265–273. Springer, Cham (2020). https://doi.org/10.1007/978-3-030-21990-1_16

8. Delibašić, B., Radovanović, S., Jovanović, M., Bohanec, M., Suknović, M.: Integrating knowledge from DEX hierarchies into a logistic regression stacking model for predicting ski injuries. J. Decis. Syst. **27**(sup1), 201–208 (2018)
9. Delibašić, B., Radovanović, S., Jovanović, M., Obradović, Z., Suknović, M.: Ski injury predictive analytics from massive ski lift transportation data. Proc. Inst. Mech. Eng. Part P: J. Sports Eng. Technol. **232**(3), 208–217 (2018)
10. Delibašić, B., Radovanović, S., Jovanović, M., Vukićević, M., Suknović, M.: An investigation of human trajectories in ski resorts. In: Trajanov, D., Bakeva, V. (eds.) ICT Innovations 2017. CCIS, vol. 778, pp. 130–139. Springer, Cham (2017). https://doi.org/10.1007/978-3-319-67597-8_13
11. Guo, W., Zhang, K., Lin, L., Huang, S., Xing, X.: Towards interrogating discriminative machine learning models. arXiv preprint arXiv:1705.08564 (2017)
12. Hume, P.A., Lorimer, A.V., Griffiths, P.C., Carlson, I., Lamont, M.: Recreational snow-sports injury risk factors and countermeasures: a meta-analysis review and Haddon matrix evaluation. Sports Med. **45**(8), 1175–1190 (2015)
13. Lundberg, S.M., Erion, G.G., Lee, S.I.: Consistent individualized feature attribution for tree ensembles. arXiv preprint arXiv:1802.03888 (2018)
14. Malasevska, I.: Explaining variation in alpine skiing frequency. Scand. J. Hosp. Tour. **18**(2), 214–224 (2018)
15. Radovanović, S., Delibašić, B., Suknović, M.: A multilabel prediction model for predicting part of the body and type of ski injury. In: Proceedings of the 4th International Conference on Decision Support System Technology – ICDSST 2018 & PROMETHEE DAYS 2018, EWG-DSS, Heraklion, Greece, p. 41 (2018)
16. Radovanović, S., Delibašić, B., Suknović, M.: Multi-task learning for ski injury predictions. In: Central European Conference on Information and Intelligent Systems, Varazdin, Croatia, pp. 215–222. Faculty of Organization and Informatics (2018)
17. Radovanovic, S., Delibasic, B., Suknovic, M., Matovic, D.: Where will the next ski injury occur? A system for visual and predictive analytics of ski injuries. Oper. Res. 1–20 (2019)
18. Ribeiro, M.T., Singh, S., Guestrin, C.: Why should I trust you?: explaining the predictions of any classifier. In: Proceedings of the 22nd ACM SIGKDD International Conference on Knowledge Discovery and Data Mining, San Francisco, USA, pp. 1135–1144. ACM (2016)
19. Ruedl, G., Benedetto, K.P., Fink, C., Bauer, R., Burtscher, M.: Factors associated with self-reported failure of binding to release among recreational skiers: an epidemiological study. Curr. Issues Sport Sci. (CISS) **1**, 003 (2016)
20. Ruedl, G., Helle, K., Tecklenburg, K., Schranz, A., Fink, C., Burtscher, M.: Factors associated with self-reported failure of binding release among ACL injured male and female recreational skiers: a catalyst to change ISO binding standards? Br. J. Sports Med. **50**(1), 37–40 (2016)
21. Ruedl, G., Kopp, M., Sommersacher, R., Woldrich, T., Burtscher, M.: Factors associated with injuries occurred on slope intersections and in snow parks compared to on-slope injuries. Accid. Anal. Prev. **50**, 1221–1225 (2013)
22. Ruedl, G., Philippe, M., Sommersacher, R., Dünnwald, T., Kopp, M., Burtscher, M.: Current incidence of accidents on Austrian ski slopes. Sportverletzung Sportschaden: Organ der Gesellschaft fur Orthopadisch-Traumatologische Sportmedizin **28**(4), 183–187 (2014)
23. Su, W., Yuan, Y., Zhu, M.: A relationship between the average precision and the area under the ROC curve. In: Proceedings of the 2015 International Conference on the Theory of Information Retrieval, Northampton, USA, pp. 349–352. ACM (2015)
24. Varmuza, K., Filzmoser, P.: Introduction to Multivariate Statistical Analysis in Chemometrics. CRC Press, Boca Raton (2016)

Performance Evaluation of Word and Sentence Embeddings for Finance Headlines Sentiment Analysis

Kostadin Mishev[1]([✉]), Ana Gjorgjevikj[1], Riste Stojanov[1], Igor Mishkovski[1], Irena Vodenska[2], Ljubomir Chitkushev[2], and Dimitar Trajanov[1]

[1] FCSE, "Ss. Cyril and Methodius" University in Skopje, Skopje, Republic of North Macedonia
kostadin.mishev@finki.ukim.mk
[2] Boston University, Boston, USA

Abstract. Nowadays, tremendous number of financial online articles are published every day. Numerous natural language processing (NLP) algorithms and methodologies have arose, not only for correct, but also for fast financial sentiment extraction. Currently, word and sentence encoders are popular topic in NLP field, due to their ability to represent them as dense vectors in a continuous real numbers space, referred to as embeddings. These low dimensional embedding vectors are appropriate for deep neural networks (DNN) inputs, and their invention boosted the performance of multiple of NLP tasks.

In this paper, we evaluate different word and sentence embeddings in combination with standard machine learning and deep-learning classifiers for financial texts sentiment extraction. Our evaluation shows the BiGRU+Attention architecture with word embedding as features, give the best score in overall evaluation.

Keywords: Sentiment analysis · Finance · Deep learning · Word embedding · Sentence embedding

1 Introduction

The stock trading in the modern world can not be imagined without technology. The one that has better information has a competitive advantage and can make smarter trade that potentially will produce a higher profit. However, obtaining the "better information" from the vast amount of continuous data, mainly originating from news and social media posts and discussions, is not naive task. There is another constraint in this equation: Every trader is seeking for the "information" that will maximize his/hers profit. The race for the right information that will provide an inside which symbols to bid or offer is what makes the stock trading exiting. And this is the place where the technology takes over. It is impossible for one person or a group of people to process all the information

© Springer Nature Switzerland AG 2019
S. Gievska and G. Madjarov (Eds.): ICT Innovations 2019, CCIS 1110, pp. 161–172, 2019.
https://doi.org/10.1007/978-3-030-33110-8_14

related to the companies and their corresponding exchange symbols in a timely matter. Especially in the high frequency trading world, where few milliseconds can change the profit into a loss, everybody tries to create a model that will increase the winning probability.

One of the pillars of the efficient market hypothesis proposes that the available information is reflected immediately in stock prices [17]. Diverse studies refer that sentiment may affect future returns [3,19], volatility and trading volume [1,19]. Driven from this incentive, the Task 5 of the SemEval 2017 challenge was introduced [9]. This particular task attracted 32 participating teams from various parts of the world, 25 of which tried to extract the sentiment from the Microblog messages, while 29 teams tried to extract the sentiment from the news headlines.

The teams that submitted a solution for the Task 5 of the SemEval 2017 challenge approach the problem with various tools and techniques, among which the most promising ones are based on hybrid models that apply machine learning, deep learning and lexicons in order to extract the sentiment from the corresponding text.

According to the study [5], successful sentiment analysis should unify the best of the natural language processing field. First, the syntax insights should be extracted through sentence boundary disambiguation, part of speech tagging, text chunking and lemmatization, and then the semantics should be understood using word sense disambiguation, concept extraction, named entity recognition, anaphora resolution and subjectivity detection. At the end is the pragmatic layer, where the sarcasm should be detected, metaphors should be understood and even then one can extract the sentiment of the sentence.

In this paper we compare the performance of multiple machine learning and deep learning approaches for financial news headlines sentiment extraction with respect to different word and sentence embeddings.

Word embedding is a technique for mapping high-dimensional sparse vectors that represent the words into a real number vector in continuous space based on the word cooccurence in large text corpora, providing significant dimensionality reduction. This representation is famous for representing the similar words together, as well as their relationships, enabling analogy detection [16]. In this paper, we compare the influence of pre-trained embeddings for the financial news headline sentiment extraction. Particularly, we use 300 dimensional word representation from Word2Vec [16], the 100 dimensional Glove [18] embedding and 300 dimensional FastText embedding [4].

Sentence embedding provides mapping of sentences into vectors of real numbers. Considering the stored information which each sentence contains, the embedding vector is larger and usually varies between 512 and 1024. In this paper we examine Universal Sentence Encoder (USE) [6] with 512 dimensional sentence representation and Language Agnostic Sentence Representation (LASER) [20], which represents the sentences with 1024 dimensional vectors.

2   Datasets

In this paper, we use two different datasets composed of financial headlines from two different sources.

Table 1. Financial Phrase Bank and SemEval2017 Task5 datasets statistics

Category	Financial Phrase Bank	SemEval2017-Task5
Neutral	2879	38
Negative	604	451
Positive	1363	653

2.1   Financial Phrase Bank

The work in [15] presents human-annotated public corpus of financial sentences extracted from English news on all OMX Helsinki listed companies. Text articles are downloaded from LexisNexis database balancing the coverage of small and large companies. It consists of approximately 5000 sentences annotated with three labels: **positive, negative** and **neutral** depending on the influence of the statement presented in the sentences to the main actor of the sentence, in this case, the company. The number of sentences per label of the dataset is given in Table 1.

2.2   Financial Headlines, SemEval 2017, Task 5

Additionally, we use dataset from the SemEval 2017 competition, task 5 named as Fine-Grained Sentiment Analysis on Financial News. It consists of around 1200 news headlines related to large companies world-wide extracted from various sources on Internet like Yahoo Finance. Each instance, in this case sentence, is labeled with a floating number ranging from -1 to 1, presenting the sentiment score. The score of 0 means neutral. Due to the consistency with previous dataset, we convert scores in nominal values: positive, negative and neutral. The conversion algorithm is presented in Eq. 1.

$$L = \begin{cases} Positive, & \text{if } score > 0 \\ Negative, & \text{if } score < 0 \\ Neutral, & \text{if } score = 0 \end{cases} \tag{1}$$

After the conversion, the number of sentences per label is presented in the Table 1.

Table 2. Merged dataset statistics

Category	Total	Training	Test
Positive	2016	1414	602
Negative	1055	762	331

2.3   Data Pre-processing

Considering the small size of the both datasets, we merge them into one. Additionally, we remove neutral sentences due to the specificity of the task, to identify positive or negative magnitude of the sentence. Next, we applied lower-casing and stemming of the words in the sentences in order to eliminate the noise which may be produced by formatting, typos and grammar additions. Most of the sentences are 10 words long, as shown in Fig. 1, which presents the distribution of the lengths of sentences in the whole dataset.

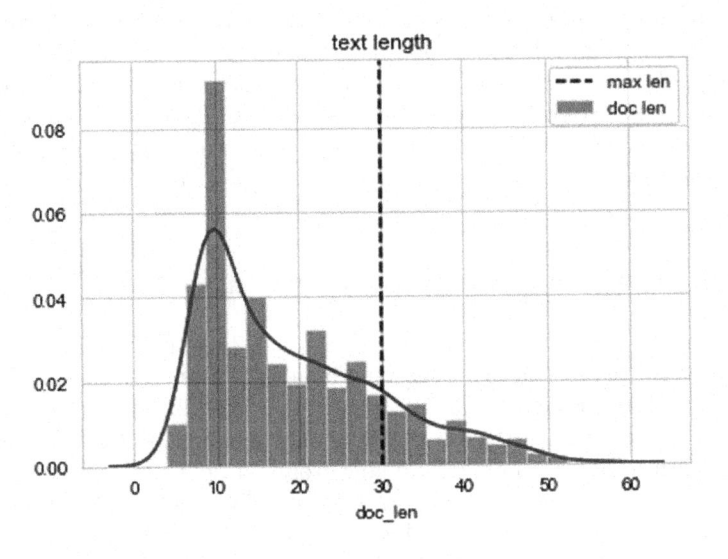

Fig. 1. Distribution of number of words per sentence

Next, we split dataset into training and test set to evaluate the models. We use 70% of the dataset as training and 30% as test set. During the training of the model, we use 10% of the training set as validation set. Number of positive and negative sentences in training and test set is given in Table 2.

3   Sentiment Analysis for Finance Text Articles

In this paper, we compare different approaches for text classification problem in finance by using deep learning methods aided by standard machine learning classifiers. The main goal is to investigate the influence of different word and sentence embeddings as inputs for these techniques. In evaluation process, we use Python programming language and we leverage the Scikit-learn library[1] for machine learning classifiers implementations, and Keras[2] for building the structure of neural network classifiers used in the proposed methodologies.

3.1   Lexicon Based Approach

Table 3. WordCount LM and VADER performance

Metric	Wordcount LMD	VADER sentiment
Accuracy	0.35	0.42
Recall	0.23	0.42
Precision	0.45	0.40
F1-score	0.30	0.41

As a starting point, we calculate the sentiment score by Wordcount Loughran-McDonald (LM) [14], as standard approach in finance, and VADER NLTK toolkit [12], as standard approach in NLP. These algorithms are based on statistical semantic analysis and they calculate the number of positive and negative words in each sentence. Word-count LM, as primary source for words polarity, uses Loughran-McDonald Dictionary [14]. VADER (Valence Aware Dictionary and sEntiment Reasoner) is a lexicon and rule-based sentiment analysis tool which is specifically attuned to sentiments expressed in social media [12]. Afterwards, the polarity of words is used to determine the polarity of sentence. If polarity of 2/3 or more negative words, the sentence is negative. If polarity of 2/3 or more positive words, the sentence is positive. The results of evaluation with test dataset is given in Table 3.

3.2   Deep Learning Methods

Our deep learning methods are based on word and sentence encoders with pre-trained models in combination with standard machine learning and deep-learning classifiers. The problem which we deal with is a binary classification problem. We evaluate each of the encoders separately. Hence, we use two separate architectures.

---

[1] Scikit-learn, https://scikit-learn.org/stable/.
[2] Keras, https://keras.io/.

Word Embedding Architectures

***Bidirectional Gated Recurrent Unit with Attention (BiGRU+
Attention).*** This model uses a variation of the Recurrent Neural Network architecture [10] based on shallow BiDirectional Gated Recurrent Unit (GRU) aided
by attention layer. We chose this architecture since the attention layer has been
proven as efficient in Machine Vision, and recently it was adopted in NLP tasks
[2], where it improves the earlier performance. Bidirectional GRU summarizes
the context of the sentences from both directions of the sequences, which in our
case acts as a sentence encoder. In [13], this model outperforms other methodologies giving best results in evaluation phase. GRU [7] is a simplified version
of Long Short Term Memory (LSTM) [11] that has been shown to outperform
LSTM in some tasks [8]. Also, in [21], the GRU units in combination with word
encoders, outperform LSTM in sentiment analysis for Russian Tweets.

The architecture that we use is presented in Fig. 2. The first layer provides the
different word embeddings that are evaluated in this scenario. For each sentence,
these embeddings are padded with empty word embedding in order to fit the
300 BiGRU units. This layer outputs a sentence encoding which is fed into the
attention layer, which learns the most "sentimental" words in the sentence i.e.
the one that influence the most for the given label. The fourth layer is a dense
layer which is feed in a dropout layer in order to prevent over-fitting. At the end,
we use the hyperbolic tangent in order to predict the label.

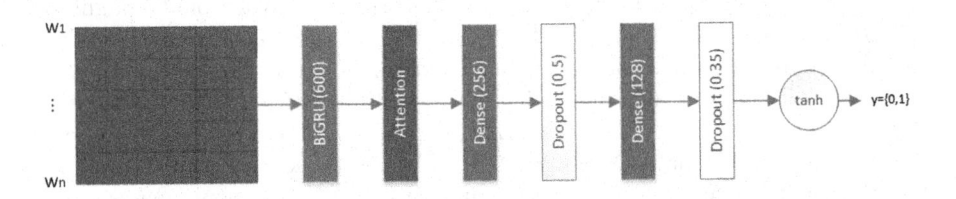

Fig. 2. BiGRU+Attention with Word embedding architecture

Convolution Neural Network (CNN). Convolution Neural Network (CNN)
can extract the most informative n-grams in the sentence which affect the classification and polarity of the sentence. Our CNN network with word embedding
layer consists of two 1-dimensional convolution layers with 64 filters separated
by max-pooling layer with pool size 2. Fifth layer is Global Max Pooling layer,
which extracts the max vector over the steps, which outputs a constant dimension for the temporal data. Next, we have Dropout layer with probability 0.5
followed by dense layer with 32 neurons and sigmoid dense layer with 2 neurons.
The presented architecture is visualized in Fig. 3.

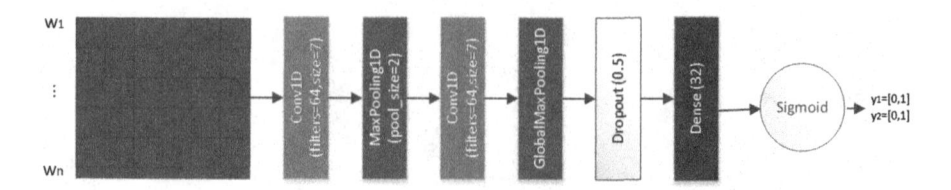

Fig. 3. CNN Word embedding architecture

Sentence Embedding Architectures

Bidirectional Gated Recurrent Unit (BiGRU). The architecture of this network is presented in Fig. 4. The only difference with word embedding BiGRU architecture is that it accepts a sentence embedding, and does not have the attention layer. The output of the network is sigmoid function which produces two outputs, prediction probabilities for two classes.

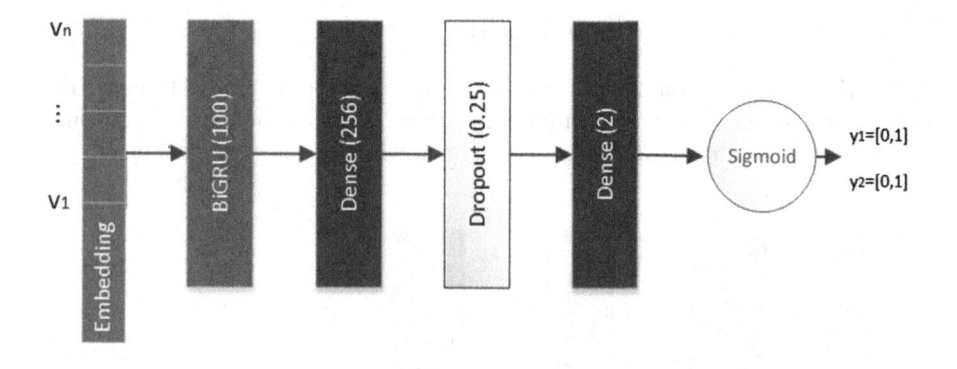

Fig. 4. BiGRU architecture for sentence embedding

Convolution Neural Network (CNN). In this model we stack 2 stacks of convolution layers with 512 filters and filter size 7. In the middle, we set Max Pooling layer with pool size 2. To the output of the seconds convolution layer, we set Dropout layer in order to reduce the over-fitting of the network. Next, we put dense layer with ReLU activation function and 32 neurons and lastly, we set dense layer with sigmoid activation function and 2 neurons which, as output, give probability values for predicting the appropriate class. As model optimizer, we use stochastic gradient descent and we use categorical crossentropy as loss function. The details of the implementation are given in the Fig. 5.

Multi-layer Perceptrons (MLP). Additionally, we test multi-layer perceptron deep learning architecture in text classification. we fed sentence embedding received by embedding layer into four hidden dense layers having 1024, 512,

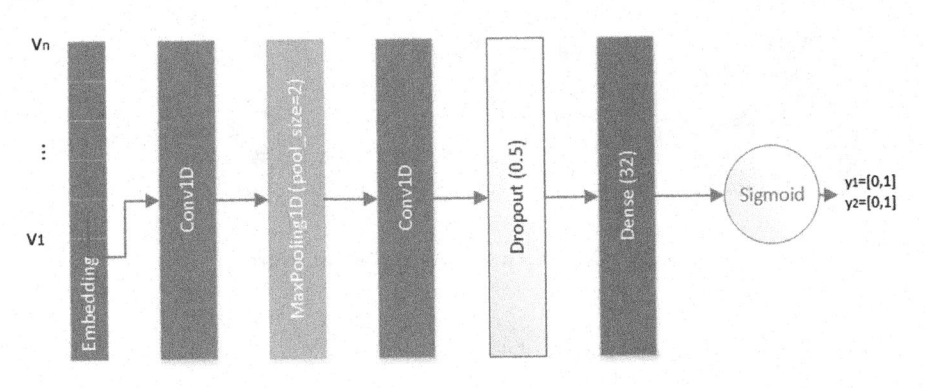

Fig. 5. CNN architecture for sentence embedding

64 and 2 neurons respectively. For each sentence embedding vector representation $x_i = [x_{i1}, x_{i2}, ..., x_{iT}]$ of sentence m_i, each neuron of the hidden network calculates a vector h_j defined by the follow equation.

$$h_{ij} = ReLU(W_{ij}x_i + b_j) \tag{2}$$

where W_{ij} is the weight matrix and b_j is the bias vector of the layer j. The output of the network is a softmax dense layer which produces two outputs, prediction probabilities for two classes. The architecture is given in Fig. 6

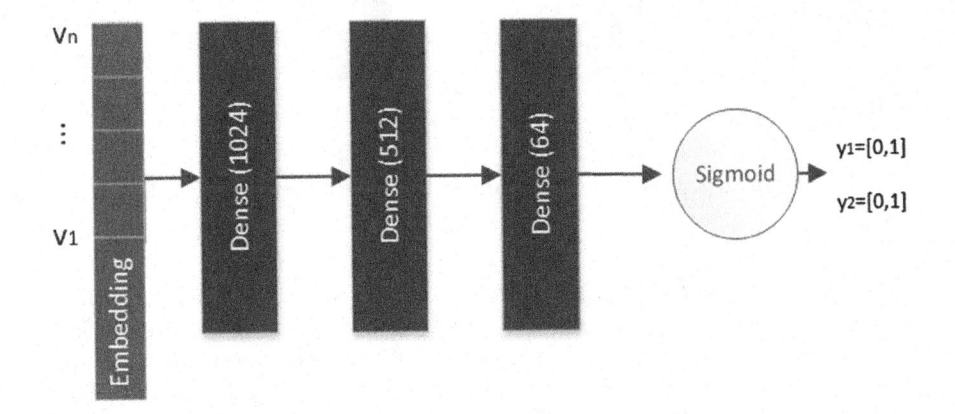

Fig. 6. MLP architecture for sentence embedding

4   Evaluation

Tables 3–5 present results of the evaluation on sentiment classification. Table 3 presents the results from lexical representation with ML methods. Table 4 presents the results by using word encoders with deep-learning methods. Table 5 presents results obtained in binary classification aided by sentence encoders with ML and DL methods. All tables show macro-averaged F1-score of negative and positive classes, used as quality measure.

LASER, as sentence embedding, with BiGRU architecture outperforms other sentence embeddings. Meanwhile, Glove with BiGRU+Attention achieved best results in sentiment identification among word encoders.

Confusion matrices in Fig. 7 present the results for positive and negative class predictivity for the best models from word and sentence embeddings respectively used over the test-set. In both models, we achieve better results in positive class prediction.

Table 4. Word Encoder F1 score

Encoder	Word2Vec	Glove	FastText
BiGRU+Attention	0.885	**0.893**	0.890
CNN	0.857	0.870	0.865

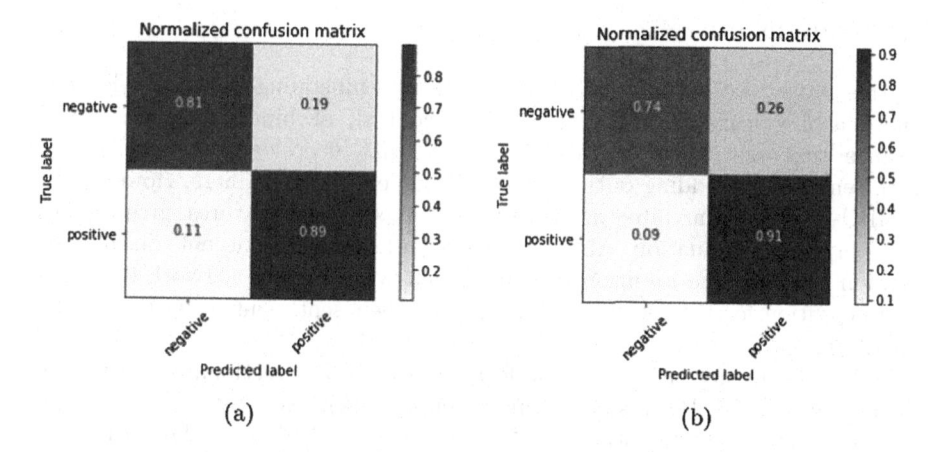

Fig. 7. Confusion matrix of predicted labels in test set (a) Glove (b) LASER

Table 5. Sentence Encoder F1 score

Encoder	USE	LASER
BiGRU	0.846	**0.889**
CNN	0.784	0.812
MLP	**0.864**	0.869
Extreme Gradient Boosting (max_depth = 80, n_estimators = 120, lr = 0.1)	0.828	0.82
Extra Random Forest (n_estimators = 100, max _depth = 2)	0.838	0.82
GradientBoosting (n_estimators = 200, max_depth = 3)	0.828	0.84
SVC Gamma (gamma = 2, C = 1)	0.851	0.82
Stochastic Gradient Descent (loss = "hinge", penalty = "l2", max_iter = 5)	0.750	0.82
Passive Aggressive (max_iter = 1000, random_state = 0)	0.770	0.82
Decision Tree Classifier	0.729	0.72
K-Nearest Neighbour (n_neighbors = 5)	0.787	0.67
Logistic Regression	0.827	0.72
Naive Bayes	0.721	0.66
Random Forest	0.789	0.56
SVC Linear (C = 0.025)	0.780	0.39

5   Conclusion

In this paper, we evaluate word and sentence embedding with deep-learning and machine learning architectures for the task of binary classification of finance text sentiment. The results show that deep-learning architectures with sentence embedding outperform machine-learning classifiers. However, the BiGRU+Attention architecture with word embedding as features, give the best score in overall evaluation. Also, the attention layer does a decent job in classification process. The intuition for this is that we train this network to learn a representation for the sentences that embeds their sentiment using the BiGRU and Attention layers.

LASER sentence encoder outperforms the USE. Bidirectional GRU in combination with LASER gives best result when using sentence embeddings as input for the classifier. On the other hand, for the USE embeddings, the MLP architecture showed best results.

6   Future Work

In order to improve the performances of evaluation, we need a larger dataset. As next step, we plan to enrich our dataset with additional news headlines from Thompson Reuters News database. On the other side, we can use unsupervised approach for data augmentation as presented in [22]. Also, we should

evaluate combinations of different embeddings by using deep neural networks with concatenation layer. Additionally, LASER and FastText embeddings are language-agnostic, so we plan to measure their performances with texts written in different languages.

References

1. Antweiler, W., Frank, M.Z.: Is all that talk just noise? The information content of internet stock message boards. J. Finan. **59**(3), 1259–1294 (2004)
2. Ba, J., Mnih, V., Kavukcuoglu, K.: Multiple object recognition with visual attention. arXiv preprint arXiv:1412.7755 (2014)
3. Baker, M., Wurgler, J.: Investor sentiment in the stock market. J. Econ. Perspect. **21**(2), 129–152 (2007)
4. Bojanowski, P., Grave, E., Joulin, A., Mikolov, T.: Enriching word vectors with subword information. Trans. Assoc. Comput. Linguist. **5**, 135–146 (2017)
5. Cambria, E., Poria, S., Gelbukh, A., Thelwall, M.: Sentiment analysis is a big suitcase. IEEE Intell. Syst. **32**(6), 74–80 (2017)
6. Cer, D., et al.: Universal sentence encoder. arXiv preprint arXiv:1803.11175 (2018)
7. Chung, J., Gülçehre, Ç., Cho, K., Bengio, Y.: Empirical evaluation of gated recurrent neural networks on sequence modeling. CoRR abs/1412.3555 (2014). http://arxiv.org/abs/1412.3555
8. Chung, J., Gulcehre, C., Cho, K., Bengio, Y.: Empirical evaluation of gated recurrent neural networks on sequence modeling. arXiv preprint arXiv:1412.3555 (2014)
9. Cortis, K., et al.: SemEval-2017 Task 5: fine-grained sentiment analysis on financial microblogs and news. In: Proceedings of the 11th International Workshop on Semantic Evaluation (SemEval-2017), pp. 519–535 (2017)
10. Elman, J.L.: Finding structure in time. Cogn. Sci. **14**(2), 179–211 (1990)
11. Hochreiter, S., Schmidhuber, J.: LSTM can solve hard long time lag problems. In: Advances in Neural Information Processing Systems, pp. 473–479 (1997)
12. Hutto, C.J., Gilbert, E.: Vader: a parsimonious rule-based model for sentiment analysis of social media text. In: Eighth International AAAI Conference On Weblogs and Social Media (2014)
13. Kar, S., Maharjan, S., Solorio, T.: RiTUAL-UH at SemEval-2017 Task 5: sentiment analysis on financial data using neural networks. In: Proceedings of the 11th International Workshop on Semantic Evaluation (SemEval-2017), pp. 877–882 (2017)
14. Loughran, T., McDonald, B.: When is a liability not a liability? Textual analysis, dictionaries, and 10-Ks. J. Finan. **66**(1), 35–65 (2011)
15. Malo, P., Sinha, A., Takala, P., Ahlgren, O., Lappalainen, I.: Learning the roles of directional expressions and domain concepts in financial news analysis. In: 13th International Conference on Data Mining Workshops, pp. 945–954. IEEE (2013)
16. Mikolov, T., Sutskever, I., Chen, K., Corrado, G.S., Dean, J.: Distributed representations of words and phrases and their compositionality. In: Advances in Neural Information Processing Systems, pp. 3111–3119 (2013)
17. Mishkin, S., Eakins, G.: Financial Markets and Institutions. Prentice Hall, Boston (2012)
18. Pennington, J., Socher, R., Manning, C.: Glove: global vectors for word representation. In: Proceedings of the 2014 Conference on Empirical Methods in Natural Language Processing (EMNLP), pp. 1532–1543 (2014)

19. Sabherwal, S., Sarkar, S.K., Zhang, Y.: Do internet stock message boards influence trading? Evidence from heavily discussed stocks with no fundamental news. J. Bus. Finan. Account. **38**(9–10), 1209–1237 (2011)
20. Schwenk, H., Douze, M.: Learning joint multilingual sentence representations with neural machine translation. arXiv preprint arXiv:1704.04154 (2017)
21. Trofimovich, J.: Comparison of neural network architectures for sentiment analysis of Russian Tweets. In: Computational Linguistics and Intellectual Technologies: Proceedings of the International Conference Dialogue (2016)
22. Xie, Q., Dai, Z., Hovy, E., Luong, M.T., Le, Q.V.: Unsupervised data augmentation. arXiv preprint arXiv:1904.12848 (2019)

A Hybrid Model for Financial Portfolio Optimization Based on LS-SVM and a Clustering Algorithm

Ivana P. Marković[1], Jelena Z. Stanković[1], Miloš B. Stojanović[2], and Jovica M. Stanković[1(✉)]

[1] Faculty of Economics, University of Niš,
Trg kralja Aleksandra Ujedinitelja 11, 18000 Niš, Serbia
{ivana.markovic, jelenas,
jovica.stankovic}@eknfak.ni.ac.rs
[2] College of Applied Technical Sciences Niš,
Aleksandra Medvedeva 20, 18000 Niš, Serbia
milos.stojanovic@vtsnis.edu.rs

Abstract. An investment decision is one of the most important financial decisions. With the aim of optimizing investment in securities from the aspect of return and risk, investors usually diversify their portfolio securities. This paper presents a hybrid model for portfolio optimization, which consist of two steps. The first step predicts future returns on the shares, and the second step, by applying hierarchical clustering algorithm, identifies various groups of shares. The test results indicate that the suggested model is suitable for optimization of a financial portfolio as a hybrid model based on selected shares, which if included in the portfolio, enable the diversification of risk.

Keywords: Portfolio optimization · Recursive prediction · Dynamic time warping · Hierarchical clustering

1 Introduction

Investment in the capital market represents a complex process of allocating free financial resources with the aim of achieving a certain return. The highly volatile nature of capital markets prevents the accurate analysis, prediction and calculation of the investment return and risk, making these investments the riskiest. Portfolio optimization represents one of the most important aspects of making investment decisions, which is based on the possibility of the successful prediction of the relationship between return and risk in the future. Modern portfolio theory [1] is the first quantitative framework for the optimal diversification of investment decisions, which is widely being used even today. According to Markowitz's approach, investment optimization model is based only on the values of the expected return and risk, while it is assumed that a rational investor prefers more effective portfolios, which provide maximal return for acceptable level of risk. The development of the stock market enhances the possibilities for investment, but increases the complexity of the model. In

© Springer Nature Switzerland AG 2019
S. Gievska and G. Madjarov (Eds.): ICT Innovations 2019, CCIS 1110, pp. 173–186, 2019.
https://doi.org/10.1007/978-3-030-33110-8_15

cases when a large number of financial instruments are being considered, high dimensionality can prevent the precise evaluation of a correlation structure and risk. Thus, the application of Markowitz's model in the optimization of large portfolios can result in the allocation of financial resources to suboptimal investment alternatives.

The problem of diversification when creating effective portfolios in comparison to more conventional methods could also be solved by the application of numerous alternative methods. In numerous studies, machine learning algorithms have proved to be quite effective in the creation of effective portfolios. At the same time, the most frequently used approaches include artificial neural networks [2, 3], genetic algorithms [4, 5], and cluster analysis [6–9].

Cluster analysis represents a technique of unsupervised machine learning which attempts to establish previously unknown and useful connections among the data, whereby different insight into the structure of the data is obtained. A special type of clustering is the clustering of time series, which are classified as dynamic data since their characteristics can change over an observed period of time. Additionally, clustering of financial time series is an especially challenging task, taking into consideration the fact that the financial market is a complex, evolving and dynamic system whose behavior is pronouncedly non-linear.

The existing portfolio optimization systems usually focus on several aspects: improvement in the clustering approaches [8, 10], improvement in the used metrics for the allocation of clustered data [11, 12], or methods for optimization of asset selection from the created clusters [13]. The problem of improving clustering algorithms based on the prediction of the available data set for clustering, however, has so far not been studied in sufficient detail, particularly in the field of financial time series data. In this paper, as a contribution, we propose a two-step methodology for the improvement of the performance of the clustering algorithm based on newly available data obtained by using machine learning algorithms for prediction. To our best knowledge, the proposed hybrid approach for optimization of financial portfolios has not been used so far.

In this paper LS-SVMs will be used as a first step in the proposed methodology to create a prediction model. The problem of prediction of asset returns is modeled with a recursive time series prediction strategy. In the second step of the proposed hybrid methodology for portfolio optimization, hierarchical clustering is used.

It is important to point out that any regression machine learning technique is suitable for the application of the proposed hybrid portfolio optimization methodology in combination with any other clustering algorithm with adequate metrics. The proposed algorithm offers a systematic approach for financial portfolio optimization based on predicted data.

The paper is organized as follows: the second part describes the proposed methodology. The third part presents some theoretical basics needed to understand the proposed methodology. The fourth part represents the simulation framework and provides an overview of the results of the cluster analysis and the results of the portfolio optimization. In the final part, some of the conclusions and directions for further research are presented.

2 Proposed Methodology

In this section a hybrid methodology for portfolio optimization is presented, which is based on a recursive time-series prediction strategy and a hierarchical clustering algorithm.

We will first describe the recursive time-series prediction strategy. It is the most intuitive strategy of long-term prediction [14–16], considering that it relies on the already predicted values instead of familiar ones, which in the given moments were not available, for the prediction of future values. In the case of recursive strategies, prediction for one step ahead is done:

$$\hat{y}_{t+1} = f(y_t, y_{t-1}, \ldots, y_{t-d+1}) \tag{1}$$

The number of previous values of the time series based on which the prediction of the following value is made is defined by the regressor size d, while f stands for the prediction model. To predict the following step, we use the same model, with the same group of parameters:

$$\hat{y}_{t+2} = f(\hat{y}_{t+1}, y_t, y_{t-1} \ldots, y_{t-d+2}) \tag{2}$$

and finally, to predict the H-th step, where H denotes the prediction horizon size, $\hat{y}_{t+H}$ is predicted using the same model:

$$\hat{y}_{t+H} = f(\hat{y}_{t+H-1}, \hat{y}_{t+H-2}, \ldots, \hat{y}_{t-d+H}) \tag{3}$$

In (2) the predicted value was used for $\hat{y}_{t+1}$, instead of its correct value, which at the given time was not available. To predict all the H steps, the values from $\hat{y}_{t+2}$ to $\hat{y}_{t+H}$ are predicted recursively, and with each step, the number of predicted values in the regressor increase. The moment when the prediction horizon size H becomes equal to or greater than the regressor size d, in the inputs to the model we find only the values from previous steps.

The advantage of the recursive strategy can be seen in the fact that it is necessary to form only one prediction model, that is, it is necessary to optimize only one group of parameters for the prediction of all the steps within the horizon H. The drawbacks are seen in the accumulation and propagation of errors through the prediction steps, which are especially prominent with the increase in the size of the prediction horizon.

In the proposed hybrid methodology, after the initial recursive time-series prediction model is used and new predicted data becomes available, we suggested using a hierarchical clustering algorithm over the predicted data in order to optimize the financial portfolio. The proposed algorithm can be seen in Fig. 1.

Figure 1 presents the general procedure for the prediction of M actions through a recursive strategy with d lags in each training vector.

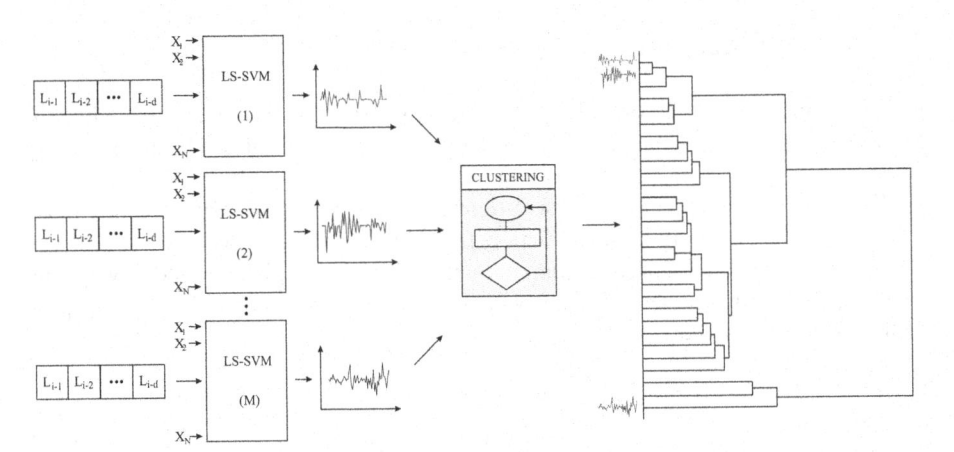

Fig. 1. Proposed hybrid methodology

The order of the steps shown in Fig. 1 can be described in the following algorithm.
Algorithm 1. The hybrid model for financial portfolio optimization

1. Initialization
   From the time series values form an initial training set S = {X, Y} with pairs from
   the training set $\{x_k, y_k\}, k = 1, \ldots N$, where N denotes the size of the initial training
   set, that is, the overall number of input-output pairs from X and Y. In the case of the
   prediction of time-series, training vectors $x_k \in X$ with the inputs $x_k \in R^n$ consist of
   lagged input variables, while $y_k \in R$ represent the target values of the time series,
   $y_k \in R$. From the last d values of the time series form the initial test vector
   $x_t = \{y_N, y_{N-1}, \ldots y_{N-d+1}\}$.
2. On the basis of this model, form and preserve the prediction for the following step
   in the time series in the vector $y_t, t = 1, \ldots, H$.
3. Update vector x_t by shifting it one place, and adding the prediction formed in step 3,
   in accordance with the recursive prediction strategy.
4. Repeat steps **2** and **3** H times, where H represents the size of the prediction horizon,
   that is, the number of steps that need to be predicted.
5. Repeat the procedure for each share which could potentially take part in the portfolio creation.
6. Following the initial prediction over the obtained values of time series, perform the
   clustering.
7. Based on the results obtained in the cluster analysis, by applying an optimization
   strategy perform the shares selection for each cluster.

If the predicted values were to be used as input, an effect may be expected on the
level of accuracy of the following prediction; therefore, a well-trained model is of vital
importance for the prevention of error propagation. The presented methodology was
tested in the experimental framework which is described in detail in section four.

3 Theoretical Framework

In this section, a brief review of LS-SVMs is given, which are used to train the forecasting models in the experiments. A theoretical basis is also needed to understand the used clustering algorithm.

3.1 Least Squares Support Vector Machines for Regression

Least squares support vector machines, as a reformulation of SVMs, are commonly used for function estimation and for solving non-linear regression problems [17]. The regression model in primal weight space is expressed in:

$$y(x) = \omega^T \phi(x) + b \tag{4}$$

where ω represents the weight vector, b represents a bias term and ϕ is a function which maps the input space into a higher-dimensional feature space.

LS-SVM formulates the optimization problem in the primal space defined by:

$$\min_{\omega, b, e} J_p(\omega, e) = \frac{1}{2}\omega^T \omega + \frac{1}{2}\gamma \sum_{k=1}^{N} e_k^2 \tag{5}$$

subject to equality constrains expressed as:

$$y_k = \omega^T \phi(x_k) + b + e_k, k = 1, \ldots, N \tag{6}$$

while e_k represents error variables, γ is a regularization parameter which gives relative weight to errors and should be optimized by the user.

Solving this optimization problem in dual space leads to obtaining α_k and b in the solution represented as:

$$y(x) = \sum_{k=1}^{N} \alpha_k K(x, x_k) + b \tag{7}$$

The dot product $K(x, x_k)$ represents a kernel function while α_k are Lagrange multipliers. When using a radial basis function (RBF) defined by:

$$k(x, x_k) = e^{-\frac{\|x-x_k\|^2}{\sigma^2}} \tag{8}$$

the optimal parameter combination (γ, σ) should be established, where γ denotes the regularization parameter and σ is a kernel parameter. For this purpose, a grid-search algorithm in combination with k-fold cross-validation is a commonly used method [18].

3.2 Hierarchical Clustering

In hierarchical clustering, the clusters are determined by applying an agglomerative or divisive algorithm. Agglomerative clustering uses a bottom-up approach to perform hierarchical clustering, so that each observation is primarily found in its special cluster,

and then the clusters are combined into larger clusters until finally all the observations belong to the same cluster or until a stopping criterion is satisfied. A divisive algorithm generates clusters following the opposite procedure, that is, in a top-down manner, beginning with one cluster which contains all the observations and then later determining the subclusters.

In this paper we used complete linkage hierarchical clustering, which uses linkage criteria to determine the metric which is used when grouping clusters. The complete linkage algorithm calculated the maximum distance between observations of pairs of clusters.

A similarity measure is of essential importance for each clustering algorithm, but similarity measures on time series data are much more difficult to determine than on constant data, because of their own continuous nature. This is why the similarity measure is, as a rule, carried out in an approximate manner [19]. When clustering time series, various types of similarity measures are used, considering that each measure reflects a different type of similarity among time series data.

The shape-based measures as the Euclidean distance and Dynamic Time Warping (DTW) [20], which find similarities in the time series in the domain of time and shape, are widely used for clustering times series data [19].

The DTW aligns two time series., $Q = q_1, q_2, \ldots, q_n$ and $P = p_1, p_2, \ldots, p_m$, using the matrix M_{nxm}, in order to minimize their difference. The element (i, j) of the matrix M represents the distance $d = (q_i, p_j)$ between two points q_i and p_j, where in order to calculate the distance, various methods can be used. In this paper we used a standard Euclidean distance. The warping path, denoted by $W, W = w_1, w_2, \ldots w_k, \ldots w_K$ while $max(m, n) \leq K \leq m + n - 1$, actually represents a group of matrix elements which define the mapping between Q and P, with the k-th element $w_k = (i_k, j_k)$. The wrapping path needs to satisfy three conditions: the continuity condition, the boundary condition and the monotonicity condition [19]. The optimal path is determined by applying dynamic programming in order to find the wrapping path W which minimizes the wrapping cost, that is, minimizes the distance between the two time series:

$$DTW(Q, P) = \min_w \left[\sum_{k=1}^{K} d(w_k) \right] \tag{9}$$

where $d(w_k) = (q_{i_k}, p_{i_k}) = (q_{i_k} - p_{i_k})^2$. The complete derivations can be found in [21].

4   Experimental Results and Discussion

This section presents the experimental results and discussion of applying the proposed approach to datasets taken from the regulated market of the Belgrade Stock Exchange.

Stock exchange indexes represent instruments used to monitor the conjunction on a particular segment of the financial market. Currently, the following indexes are calculated on the Belgrade stock exchange: BELEX15 and BELEX*line*. The values of these indexes are determined by the prices of the most liquid shares which are continuously being traded on the regulated market of the Belgrade Stock Exchange. These

indexes show the aggregate change in the price of securities on the financial market, which have previously satisfied the criteria for inclusion in the index basket.

Stocks are traded on the Belgrade Stock Exchange in both the regulated in unregulated market. On the regulated market, we can distinguish between Prime Listing, Standard Listing and the Open Market. The conditions for Prime Listing and Standard Listing were met only by the shares of seven companies, while the conditions for the Open Market were met by 25 issuers of shares. An adequate evaluation of the basic characteristics of financial time series requires a certain duration of the financial time series, which in the opinion of analysts reduced the number of available stocks to 29.

Taking into consideration the required conditions which the companies issuing the shares need to meet in order for them to be included in these lists, the starting point in this study was the assumption that they are the most liquid securities on the Belgrade Stock Exchange. Considering that they are continually being traded on the stock exchange, the time series of the data are continuous, which makes financial modelling simpler.

The data used in this paper were taken from the website of the Belgrade Stock Exchange (www.belex.rs). The training set of the series of values of individual shares, include records from January 1, 2008 to December 31, 2017. This training set was used to train the LS-SVM model. The test set for recursive prediction includes 240 days of trading, beginning with January 1, 2018. On the obtained predicted values, in the second step included in the proposed methodology, the algorithm of hierarchical clustering was applied.

The LS-SVM based recursive prediction model in this study was created based on the modification of the approach proposed in [22, 23], and consequently, through a series of experiments on time series data, we have chosen 15 lag values as input features for the recursive prediction model. The learning phase of LS-SVMs involves the optimal selection of kernel parameters, in this case σ, and the regularization parameter γ. A good choice of these parameters is essential for the estimator performance. In our experiments, we used 10-fold cross-validations in combination with grid searches for the selection of these parameters.

After training the LS-SVM, the local LS-SVM model is then employed for the time-series prediction. The whole process is repeated for each of the 29 available assets in the obtained data set by employing a recursive prediction strategy. We have used it for recursive prediction of the next 60 days for 4 time horizons for each of the 29 available assets, a total of 240 days of trading which make up approximately one year of trading per asset.

For each of the shares from the studied set, first one day ahead prediction is performed: $r(t + 1) = LS\text{-}SVM(r(t), r(t - 1), \ldots, r(t - 15))$, then to predict the asset return for the next day, the same model is used: $r(t + 2) = LS\text{-}SVM(r(t + 1), r(t), \ldots, r(t - 14))$, and for the last day: $r(t + 60) = LS\text{-}SVM(r(t + 59), r(t + 58), \ldots, r(t + 45))$.

The prediction quality was evaluated using the mean squared error (MSE) and RMSE:

$$MSE = \frac{1}{H} \sum_{i=1}^{H} (y_i - \hat{y}_i)^2 \tag{10}$$

$$RMSE = \sqrt{\frac{1}{H} \sum_{i=1}^{H} (y_i - \hat{y}_i)^2} \tag{11}$$

where y_i and $\hat{y}_i$ are the real and the predicted value of the time-series in the ith prediction step and H is the size of prediction horizon.

Figures 2 and 3 present the values of the MSE and RMSE errors obtained while predicting each of the shares in the studied subquarters.

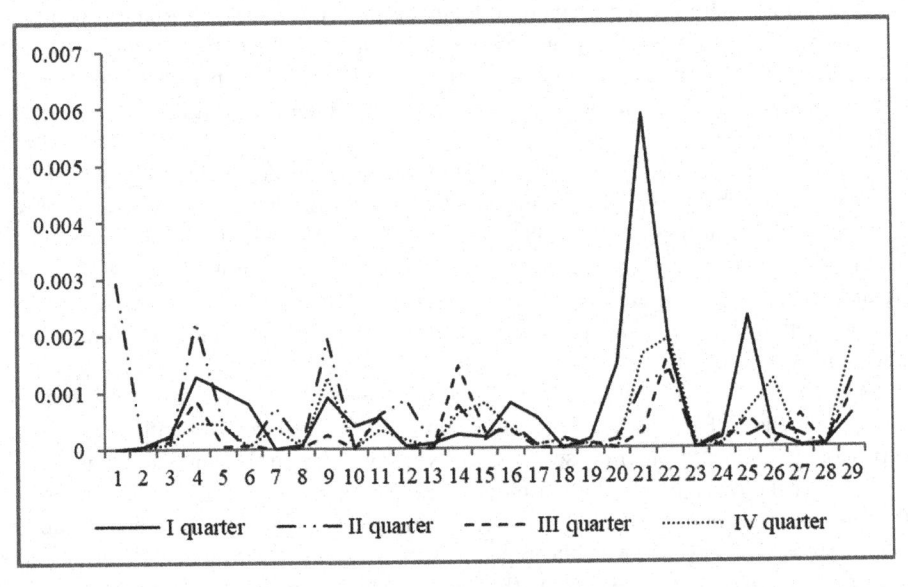

Fig. 2. MSE errors by quarters

Based on the values of the presented errors, we can note that the proposed recursive prediction model is exceptionally precise, bearing in mind the studied time horizon.

The second step uses the clustering algorithm for financial time series clustering. We tested the potential of clustering in portfolio optimization using a complete linkage (CL) agglomerative clustering algorithm by applying the DTW metric.

Figure 4 shows the results of the application of the DTW metric on the studied group for two shares, EPEN (6) and SJPT (22), in quarter IV.

It is known that the performance of each clustering algorithm is affected by the number of created clusters. In this paper we selected the optimal number of clusters using the Calinski-Harabaz index [24]. Considering that the aim of the paper is the portfolio optimization of the BELEX15 index which is made up of at most 15 shares, the Calinski–Harabaz index was set up to determine the optimal number of clusters for the studied shares with the limitation that the maximal number can be 15.

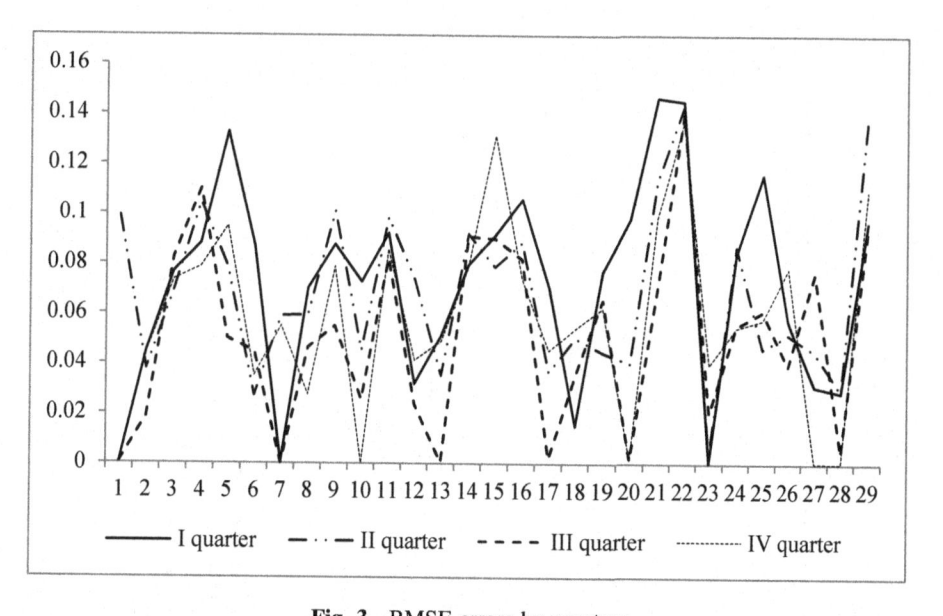

Fig. 3. RMSE errors by quarters

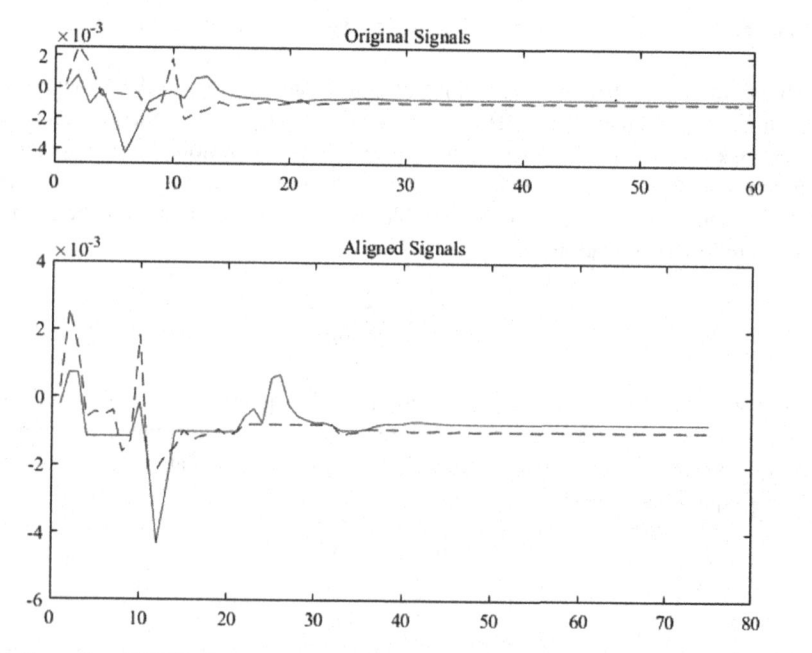

Fig. 4. Example of DTW alignment between EPEN (6) and SJPT (22) shares over a three-month period

Figure 5 shows the dendrogram of hierarchical clustering by using a complete-linkage method on the predicted data set.

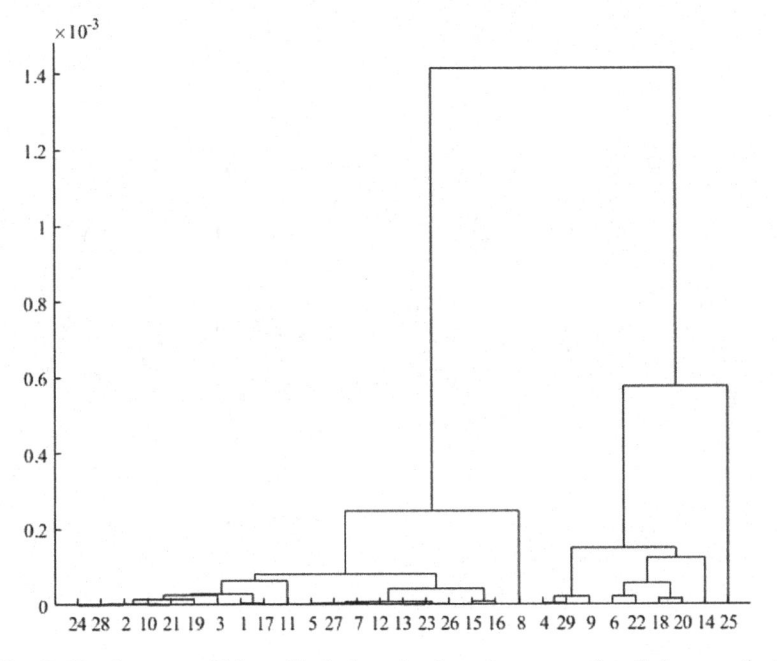

Fig. 5. Dendrogram of hierarchical clustering by using a complete linkage method

Afterwards, we distributed assets inside each cluster relying on the Omega ratio as the optimization strategy for asset allocation. The Omega ratio allows comparisons between risk and return at different threshold levels for various asset choices [25]. Assuming that $F(x)$ is a cumulative distribution function of the rate of return on investment and rm is the minimum acceptable return, the Omega ratio can be calculated using the following equation:

$$\Omega_t = \frac{\int_{r_m}^{\infty}(1 - F(x))dx}{\int_{-\infty}^{r} F(x)dx} \tag{12}$$

As a general measure for the evaluation of the quality of the obtained cluster analysis, in this paper we used three different economic performance measures: the return r_t, risk σ_t and the Sharpe ratio SR_t, which are calculated according to the following equations:

$$r_t = logCP_t - logCP_{t-1} \tag{13}$$

$$\sigma_t = \sqrt{\frac{\sum_{t=1}^{n}(r_t - \bar{r})^2}{n - 1}}. \tag{14}$$

$$SR_t = \frac{r_t}{\sigma_t} \tag{15}$$

where CP represents the closing price at t, $t = 1, 2, ..., n$.

In order to study all the aspects of behavior of the proposed hybrid model for portfolio optimization, multiple testings of the obtained results were carried out. First, on each of the prediction horizons which are represented by a fixed quarter size of 60 days, clustering was performed, followed by the optimization of the portfolio in these studied periods. The results are presented in Table 1. Finally, the TOTAL index was created based on the obtained quarter projections with the rebalancing of the portfolio, and the results of the evaluation parameters are also shown in Table 2. All of the results obtained by the portfolio optimization based on applying the proposed hybrid methodology are compared to the values of the market portfolios – the BELEX15 and BELEX*line* indexes, during the same monitored test period.

Table 1. Performance of the constructed portfolio and market portfolios (indexes) by quarters in 2018

Obtained period	Evaluation measure	Portfolio		
		CL	BELEX15	BELEX*line*
I	Return	0.022	−0.023	−0.061
	Risk	0.127	0.042	0.040
	Sharpe	0.173	−0.534	−1.539
	Omega	1.123	0.836	0.586
II	Return	−0.014	−0.013	−0.009
	Risk	0.036	0.026	0.025
	Sharpe	−0.393	−0.496	−0.346
	Omega	0.857	0.847	0.888
III	Return	0.012	−0.009	−0.011
	Risk	0.049	0.043	0.034
	Sharpe	0.246	−0.199	−0.334
	Omega	1.096	0.931	0.888
IV	Return	0.026	0.047	0.036
	Risk	0.040	0.062	0.036
	Sharpe	0.657	0.749	1.001
	Omega	1.266	1.287	1.397

Table 1 shows the results of the quarter clustering and the obtained portfolios. Each quarter is uniformly presented by 60 days of trading and is noted in the table using the numbers I–IV. The results obtained in the first test period, which includes the first quarter of 2018, indicate that the CL algorithm allows the creation of a portfolio which is characterized by a significantly greater risk, but with a more favorable relationship between return and risk compared to the market portfolios. The complete linkage method is used to create a portfolio which realizes the best results, considering that the Omega ratio of such a portfolio is 1.3 times greater than the BELEX15 index, that is, 2 times greater than the BELEXline index.

Table 2. Performance of the constructed optimal portfolio and market portfolios (indexes) during 2018

Portfolio	Return	Risk	Sharpe	Omega
CL	0.046	0.146	0.315	1.091
BELEX15	0.002	0.09	0.028	1.005
BELEX*line*	−0.045	0.068	−0.658	0.892

During the second test period, which includes the second quarter of 2018, we can note that the clustering model provides slightly worse results for the studied evaluation parameters compared to the values of the market indexes BELEX15 and BELEX*line*. However, the relationship between risk and return of such a portfolio is similar to the market indexes, which implies that the market trend in this period was in decline and not favorable for investing. Although the market maintained such a downturn trend in the third quarter of 2018, the application of the CL clustering algorithm enables the creation of a portfolio, whose performances are improved compared to those of the studied market indexes in terms of the return and risk ratio. Investing in this portfolio provides a 1.21% rate of return and a 1.2 greater Omega ratio compared to the BELEX15 and BELEX*line* indexes. Finally, in the fourth quarter of 2018 the complete linkage method enables investors to constitute a portfolio which offers an acceptable but lower risk and return tradeoff compared to the benchmark.

Table 2 pays special attention to the values of the TOTAL index which were obtained by the application of the proposed methodology with a rebalance of the portfolio at the quarter level. The TOTAL index represents a complete simulation of the market trends.

Based on the data presented in Table 2 comparing the overall performance of the clustering-based optimal portfolio to the benchmark, it can be concluded that the constructed portfolio significantly improves the features of the optimized portfolio, providing a 4.59% rate of return and the most favorable tradeoff between risk and return measured by the Sharpe (0.31) and Omega ratios (1.09).

All of the presented experiments have shown that the proposed hybrid model for financial portfolio optimization enables the diversification of risk and the constitution of a portfolio that allows investors on the Belgrade Stock Exchange to allocate their financial resources to the most favorable stocks in various market conditions.

5 Conclusion

Modern portfolio theory has provided the foundation for the development of numerous theories and concepts in finance which are still widely being used today, but has also been criticized by both the academic and professional community. Financial time series indicate a sequence of anomalies compared to the theoretical hypotheses, which require the inclusion of additional parameters with the aim of adequate risk diversification. Considering that learning from data and adapting to financial market changes are a specific and favorable advantage of machine learning methods, we have seen an increased use of these methods in financial analyses.

A practical approach to building a dynamic hybrid model for portfolio optimization is proposed. The results obtained in this study indicate the possibility of creating a portfolio by using a hybrid model in the portfolio selection task, which would result in improved performances compared to the portfolios of the market indexes in the sense of a stable diversified portfolio that shows a lower risk measured in terms of return volatility.

Future studies will focus on how the possible expansions of the group of shares which create a portfolio affect the performance of a clustering algorithm, test for adequate timing needed to rebalance a portfolio, analyze alternative strategies used to select shares from clusters included in a portfolio, and analyze the use of even more advanced clustering algorithms which might lead to further improvements in the clustering related to financial time series for portfolio optimization.

References

1. Markowitz, H.: Portfolio selection. J. Finan. **7**(1), 77–91 (1952)
2. Fernández, A., Gómez, S.: Portfolio selection using neural networks. Comput. Oper. Res. **34**(4), 1177–1191 (2007)
3. Ko, P.C., Lin, P.C.: Resource allocation neural network in portfolio selection. Expert Syst. Appl. **35**(1–2), 330–337 (2008)
4. Oh, K.J., Kim, T.Y., Min, S.: Using genetic algorithm to support portfolio optimization for index fund management. Expert Syst. Appl. **28**(2), 371–379 (2005)
5. Chang, T.J., Yang, S.C., Chang, K.J.: Portfolio optimization problems in different risk measures using genetic algorithm. Expert Syst. Appl. **36**(7), 10529–10537 (2009)
6. Nanda, S.R., Mahanty, B., Tiwari, M.K.: Clustering Indian stock market data for portfolio management. Expert Syst. Appl. **37**(12), 8793–8798 (2010)
7. Tola, V., Lillo, F., Gallegati, M., Mantegna, R.: Cluster analysis for portfolio optimization. J. Econ. Dyn. Control **32**(1), 235–258 (2008)
8. Aghabozorgi, S., The, Y.W.: Stock market co-movement assessment using a three-phase clustering method. Expert Syst. Appl. **41**(4), 1301–1314 (2014)
9. Marković, I.P., Stanković, J.M., Stanković, J.Z., Stojanović, M.B.: Financial portfolio optimization using clustering algorithms. In: 54th International Scientific Conference on Information, Communication and Energy Systems and Technologies – ICEST, 27–29 June 2019 (2019, in press)
10. Basalto, N., Bellotti, R., De Carlo, F., Facchi, P., Pascazio, S.: Clustering stock market companies via chaotic map synchronization. Physica A **345**(1–2), 196–206 (2005)
11. De Luca, G., Zuccolotto, P.: A tail dependence-based dissimilarity measure for financial time series clustering. Adv. Data Anal. Classif. **5**(4), 323–340 (2011)
12. Durante, F., Pappadà, R., Torelli, N.: Clustering of financial time series in risky scenarios. Adv. Data Anal. Classif. **8**(4), 359–376 (2014)
13. Cheong, D., Kim, Y.M., Byun, H.W., Oh, K.J., Kim, T.Y.: Using genetic algorithm to support clustering-based portfolio optimization by investor information. Appl. Soft Comput. **61**, 593–602 (2017)
14. Stojanović, M.B., Božić, M.M., Stanković, M.M., Stajić, Z.P.: A methodology for training set instance selection using mutual information in time series prediction. Neurocomputing **141**, 236–245 (2014)

15. Herrera, L.J., Pomares, H., Rojas, I., Guillen, A., Prieto, A., Valenzuela, O.: Recursive prediction for long term time series forecasting using advanced models. Neurocomputing **70**, 2870–2880 (2007)
16. Sorjamaa, A., Reyhani, J., Hao, N., Ji, Y., Lendasse, A.: Methodology for long-term prediction of time series. Neurocomputing **70**(16–18), 2861–2869 (2007)
17. Suykens, J.A.K., Van Gestel, T., De Brabanter, J., De Moor, B., Vandewalle, J.: Least Squares Support Vector Machines. World Scientific, Singapore (2002)
18. Arlot, S., Celisse, A.: A survey of cross-validation procedures for model selection. Stat. Surv. **4**, 40–79 (2010)
19. Fu, T.: A review on time series data mining. Eng. Appl. Artif. Intell. **24**(1), 164–181 (2011). https://doi.org/10.1016/j.engappai.2010.09.007
20. Berndt, D.J., Clifford, J.: Using dynamic time warping to find patterns in time series. In: KDD Workshop, vol. 10, no. 16, pp. 359–370 (1994)
21. Kruskal, J.B., Liberman, M.: The symmetric time-warping problem: from continuous to discrete. In: Kruskal, J.B., Sankoff, D. (eds.) Time Warps, String Edits, and Macro-molecules: The Theory and Practice of Sequence Comparison, pp. 125–161. CSLI Publications, Stanford (1999)
22. Gavrishckaka, V.V., Banerjee, S.: Support vector machine as an efficient framework for stock market volatility forecasting. CMS **3**(2), 147–160 (2006)
23. Gavrishckaka, V.V., Ganguli, B.S.: Volatility forecasting from multiscale and high-dimensional market data. Neurocomputing **55**(1–2), 285–305 (2003)
24. Calinski, T., Harabasz, J.: A dendrite method for cluster analysis. Commun. Stat. Theory Methods **3**(1), 1–27 (1974)
25. Keating, C., Shadwick, W.F.: A universal performance measure. J. Perform. Meas. **6**(3), 59–84 (2002)

Protein Secondary Structure Graphs as Predictors for Protein Function

Frosina Stojanovska$^{(\boxtimes)}$ and Nevena Ackovska

Faculty of Computer Science and Engineering, Ss. Cyril and Methodius University,
Skopje, Macedonia
stojanovska.frose@gmail.com, nevena.ackovska@finki.ukim.mk

Abstract. Predicting the functions of the proteins from their structure
is an active area of interest. The current trends of the secondary structure
representation use direct letter representation of the specific secondary
structure element of every amino acid in the linear sequence. Using
graph representation to represent the protein sequence provides additional information about the structural relationships within the amino
acid sequence. This study outlines the protein secondary structure with
a novel approach of representing the proteins using protein secondary
structure graph where nodes are amino acids from the protein sequence,
and the edges denote the peptide and hydrogen bonds that construct the
secondary structure. The developed model for protein function prediction Structure2Function operates on these graphs with a defined variant
of the present idea from deep learning on non-Euclidian graph-structure
data, the Graph Convolutional Networks (GCNs).

Keywords: Protein function prediction · Protein secondary
structure · Protein secondary structure graphs · Deep learning · Graph
convolutional networks

1 Introduction

Proteins are large complex molecules that play incredibly crucial roles in organisms. These molecules carry the bulk of the work in cells and are needed for
the structure, function, and regulation of tissues and organs in the organisms.
The role of the protein in a particular cell depends on the DNA sequence of
the gene synthesizing the protein, that is, from the resulting primary sequence
of amino acids. However, in organisms under given optimal conditions, proteins
are not found as a chain of amino acids but have their unique form in the three-
dimensional space.

Today there are many databases of proteins, their primary, secondary and
tertiary structures, as well as their functions, clusters, and other information [7].
The protein databases that represent the proteins with their sequence of amino
acids, such as UniProt Knowledgebase [3], grow with high rates as the protein
sequencing is getting chipper and chipper. However, annotations of proteins are

© Springer Nature Switzerland AG 2019
S. Gievska and G. Madjarov (Eds.): ICT Innovations 2019, CCIS 1110, pp. 187–201, 2019.
https://doi.org/10.1007/978-3-030-33110-8_16

still a bottleneck in the proteomics research area, so the computational protein function annotation methods are mainly developed to solve this problem.

Protein function prediction is an important application area of bioinformatics that aims to predict the Gene Ontology functions (terms) associated with the proteins. Gene Ontology (GO) represents an ontology (vocabulary) consisted of terms that describe the functions of the gene products (GO terms) and how these functions are connected between them (relationships) [4]. Currently developed methods in this field make predictions by homology searching, protein sequence analysis, protein interactions, protein structure, etc. [18]. Many proteins with similar sequences have similar functions, but there are also exceptions, such that proteins with very different sequences/structures have very similar functions [20].

Predicting the functions of the proteins from their structure is an active area of interest, closely connected to predicting the structure of the proteins. One drawback of this method is the limited number of proteins with known structures, unlike the models that utilize only the protein sequence. The solution for this problem is predicting the structure of the protein using its known amino acid sequence, which is already investigated with several existing methods. The efforts in this area are mainly divided into two sub-fields, the prediction of the secondary structure and the tertiary structure of the proteins [1,15].

As presumed from the complexity of the problem, predicting the secondary structure is a more investigative problem with methods that produce competent results. There is a clear trend of a slow but steady improvement in predictions over the past 24 years [24]. The latest techniques are approaching the theoretical upper limit of 88%–90% sufficient accuracy of predictions [23]. Hence, predicting the protein function from the secondary structure at this moment is more supportive than prediction using the ternary structure directly, mainly because of the promising solutions of using predictive models for the problem of limited structure data.

The secondary structure of the protein in currently developed models is represented as a sequence of letters indicating the affiliation of a specific amino acid from the linear sequence to one secondary structure element [10]. Consequently, the structure is described as a linear sequence of letters, similar to the primary structure - the linear sequence of amino acids. This representation of the secondary structure cannot give information about the connectivity of the amino acids, which means that we only know the particular denoted secondary element for the individual amino acid without additional information about the hydrogen bonds. This fact can be considered as a deficiency of information that is needed to extract the necessary information to predict the function by utilizing the secondary structure.

In this study, we explore the idea of generating an intelligent system capable of annotating proteins with their functions given the protein secondary structure. The secondary structure is interpreted in a novel approach, as a protein secondary structure graph. The nodes in this graph are the amino acids with their type, and these nodes are connected by two types of edges: peptide bonds and hydrogen bonds. The nodes and the peptide bonds are a representation of

the primary structure, while the addition of the hydrogen bonds supplements the information with the secondary structure. This graph comprises the knowledge of the primary structure, the peptide sequence, and also the local connection between the residues.

To predict the function using the developed protein graph, an algorithm that operates with data in the form of graphs is needed. The proposed model in this study employs the recently developed deep learning approaches for non-Euclidian graph-structured data, Graph Convolutional Networks (GCNs) [6]. The architecture of the developed method combines the GNCs with a conventional neural network, to construct a model for protein function prediction called Structure2Function.

The rest of the paper is organized as follows. Section 2 includes the details of the problem, the approach for modeling the protein structure and the proposed model for protein function annotation. Section 3 presents the experimental study and evaluation of the results and Sect. 4 concludes the study.

2 Problem Definition and Framework

The annotation of the protein is reflected as a protein classification task. The proteins are expressed with secondary structure graphs, so the proposed method for protein classification performs graph classification.

The Protein Data Bank (PDB) archive [5] is a worldwide resource for 3D structure data of biological macromolecules, which includes proteins and amino acids. Protein structure inferred through experiments is preserved in a Protein Data Bank (PDB) file. This file represents every atom of the protein molecule with relative coordinates in three-dimensional space. Hence, this file describes the protein ternary structure. The DSSP program [17] is applied to derive the information about the secondary structural elements of the folded protein. DSSP is the de facto standard for the assignment of secondary structure elements in PDB entries. The hypothesis tested here determines an intelligent system proficient for function prediction of the proteins, utilizing their secondary structure represented as a graph.

2.1 Construction of Protein Secondary Structure Graphs

The protein secondary structure can be interpreted as a graph with nodes denoting amino acids and edges outlining the chemical bonds between these amino acids. The amino acid sequence of the protein is a chain of amino acids connected by peptide covalent bonds. Therefore, one type of chemical bond between amino acids is the peptide bond. Alternatively, amino acids can form hydrogen noncovalent bond with the CO and NH groups, chemical connections that participate in the formation of secondary structure. These two types of chemical bonds are the edges in the protein graph.

The hydrogen bond that forms the secondary structure is a partially electrostatic attraction between the NH group from one residue as a donor and the

other residue considered as an acceptor with its CO group. Hence, the hydrogen bond can be modelled as a directed linkage between the donor residue to the acceptor residue. One residue can be a donor in one hydrogen bond, and acceptor in another hydrogen bond. Accordingly, one node in the defined graph model can have indegree and outdegree of the hydrogen edges equal to one. Notwithstanding, a single hydrogen atom can participate in two hydrogen bonds, rather than one, bonding called "bifurcated" hydrogen bond. Consequently, some of the nodes could have more than one incoming or outgoing edges.

The initial backbone of the protein graph is formed utilizing the peptide sequence of the protein, through the insertion of the amino acid nodes and the peptide bond edges in the graph. DSSP method is used to construct the protein secondary structure graph. The hydrogen bonds forming the secondary structure elements identified with this program are incorporated in the primary protein graph, and the outcome is the protein secondary structure graph. Figure 1 illustrates the generation procedure of the initial protein graph and the extraction of the protein secondary structure graph. This procedure is used for proteins with known tertiary structure.

2.2   Identification of Hydrogen Bonds Forming the Secondary Protein Structure

DSSP recognizes the elements of the secondary structure through hydrogen bonding patterns [17]:

- n–turn—represents a hydrogen bond between the group CO of the residue at the position i and the group NH of the residue $i + n$, where $n = 3, 4, 5$;
- bridge—signifies the hydrogen bonding between residues which are not near in the peptide sequence.

These two types of hydrogen bonding identify the possible hydrogen bonds between the amino acid residues. Six structure states are defined through the patterns of the specified hydrogen bonds: 3_{10}-helix (G), α-helix (H), π-helix (I), turn (T), β-sheet (E), and β-bridge (B).

A helical structure is defined by at least two consecutive n–turn bonds of the $i - 1$ and i residues. The ends of the helical structure are left out; that is, they are not given the appropriate state of the helical structure that they start or end. Consequently, the smallest α-helix structure is defined between the amino acid residues i and $i + 3$ if there are 4-turn hydrogen bonds of the $i - 1$ and i residues, meaning it has a length of at least 4, not counting the start and end residues. The smallest 3_{10}-helix structure between the residues i and $i + 2$ and the shortest π-helix structure between the residues i and $i + 4$ are assigned in the same way, with lengths 3 and 5 respectively. Following these rules, the subsequent n–turn H-bonding patterns define the states G, H or I. If one n-tun bond to the residue i does not have a co-occurring n–turn bond of the previous amino acid, then the residues in the range from i to $i + n$ are denoted by the state T, where n depends on the hydrogen bond itself.

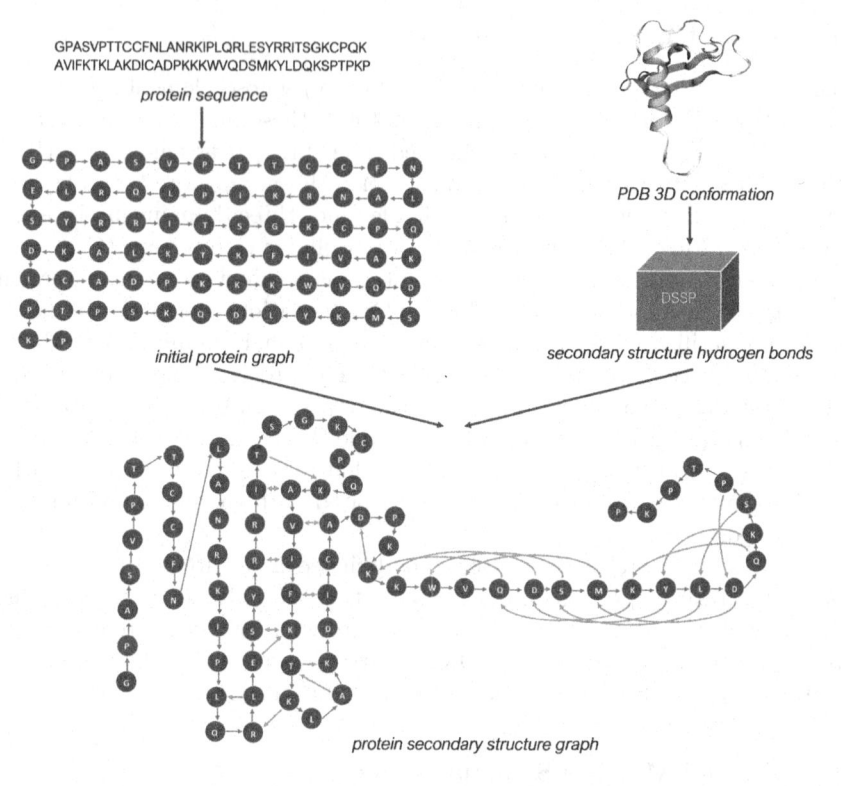

Fig. 1. Illustration of the procedure for generating the protein secondary structure graphs. The example is the PDB entry 1EOT. Circles in the protein graph represent the amino acid residues as nodes, grey arrows are the peptide bond edges, and the orange arrows are the hydrogen bond edges. (Color figure online)

The state β-sheet (E) is defined on residues with at least two consecutive bridge bonds or residues surrounded by two β-bridge hydrogen bonds. The remaining residues with inconsistent bridge hydrogen bonds obtain the β-bridge state (B). For the antiparallel β-bridge for the residues i and j it is necessary to have two connections between $i \rightarrow j$ and $j \rightarrow i$ or between $j + 1 \rightarrow i - 1$ and $i + 1 \rightarrow j - 1$. The parallel β-bridge for the residues i and j is characterized by the hydrogen bonds $j \rightarrow i - 1$ and $i + 1 \rightarrow j$ or $i \rightarrow j - 1$ and $j + 1 \rightarrow i$. The smallest β-sheet comprises two residues in each of the segments of the two strands.

The described six states are defined according to the hydrogen bonds, while the state of the bend (S) is defined geometrically according to the angles of torsion. If an amino acid residue is not found in any of the seven states, then its state is indicated by blank space and denotes a residue that is not in any of the structural elements.

2.3   Protein Functional Annotation

Gene Ontology (GO) is an ontology (vocabulary) whose elements, GO terms, describe the functions of gene products, and how these functions are related to each other [4]. This ontology is organized as a directed acyclic graph (DAG) with relations between one term on one level with one or more terms from the previous level. The ontology is mainly divided into three domains [9]: Biological Process (BP), Molecular Function (MF), and Cellular Component (CC).

One way to get functional protein annotations (current or previous versions) is through GOA annotation files. The GOA annotations annotate the protein structure, identified with their PDB identifiers and chain number, with the GO terms of the three domains BP, MF and CC. The protein annotations were filtered by removing the annotations that are not assigned by experimental methods with experimental annotation evidence code (EXP, IDA, IPI, IMP, IGI, IEP, TAS, and IC). Additionally, the annotations that have a qualifier "NOT", which indicates that the given protein is not associated with the given GO term, are also removed.

Since the structure of the ontology is designed as a directed acyclic graph (DAG), the task of protein function annotation is analyzed as a hierarchical multi-label classification. This concept indicates that for given predictions, the parent term should have a likelihood of occurrence, at least as the maximum of the probabilities of his descendant terms in the ontology.

2.4   Proposed Method: Structure2Function

The protein secondary structure graph is represented as $G = \{V, E, X\}$, where V is the set of nodes, E represents the set of edges, X is a matrix for the node content, and $N = |V|$ indicates the number of nodes. Every node has a label for the amino acid identity. There are 20 amino acids in the standard genetic code, two additional amino acids incorporated with specific processes and three other states for ambiguity (unknown amino acid or undistinguished amino acids). Therefore, every node has a one-hot encoding vector for the amino acid label with length 25. These vectors form the X matrix, where $X \in \mathbb{R}^{(N \times 25)}$.

There are two types of edges in the graph, hydrogen and peptide bond, so every edge has a label indicating the edge type. These labels are one-hot encoded; therefore, the edges have a discrete feature vector for their class. In general, the graph is directed, with no isolated nodes and no self-loops.

The architecture of the proposed method for protein classification is illustrated in Fig. 2. The model mainly consists of two major parts: graph convolutional network and fully connected neural network. The graph convolutional network maps the graph into latent representation, whereas the fully connected neural network takes the graph representation and generates the class predictions. Graph Convolutional Networks (GCNs) are already implemented in several studies for solving the task of graph classification [11,12,25].

The convolutional neural network is defined following the Message Passing Neural Networks (MPNN) framework [13]. The framework defines two phases, a

message passing phase, and a readout phase. The message passing phase updates the node representation for one node by aggregating the messages from the neighborhood of the node.

The message m_{uv} of node v from neighboring node u in our model is defined as a function from the hidden representation of node u from the previous layer h_u and the edge vector e_{uv}. This function is a neural network that takes the concatenation of h_u and e_{uv} as input and generates the message:

$$m_{uv} = \sigma(W_1[h_u, e_{uv}] + b_1) \tag{1}$$

where W_1 is a weight matrix, b_1 is a bias vector, σ denotes an activation function, and $[\cdot]$ represents concatenation. Since the graph is directed, there are two separate message channels for incoming edges m_v^{in} and outgoing edges m_v^{out}, computed as the sum of individual neighboring messages:

$$m_v^{in} = \sum_{u \in N(v)^{in}} m_{uv} \tag{2}$$

$$m_v^{out} = \sum_{u \in N(v)^{out}} m_{vu} \tag{3}$$

where $N(v)^{in}$ and $N(v)^{out}$ are the sets of neighbors from incoming and outgoing edges respectively.

The final step in the graph convolution is the node update function that updates the node hidden representation based on the m_v^{in} and m_v^{out} messages. This update function is interpreted as a neural network with h_v', m_v^{in} and m_v^{out} as inputs, where h_v' is the previous node hidden representation:

$$h_v = \sigma(W_0[h_v', m_v^{in}, m_v^{out}] + b_0) \tag{4}$$

For the first layer, the node hidden representations are the feature vectors from the matrix X. Three of the defined layers are stacked together to form a more in-depth model. Also, we use residual connections [14] between the layers to enable the transfer of the information from the previous layer:

$$H^l = GCN(H^{l-1}, A) + H^{l-1} \tag{5}$$

where l refers to the layer number, H^i is a matrix with node hidden representations, $GCN(\cdot)$ represents the graph convolutional layer, and A is the adjacency matrix.

The final step in the first part of the model is the accumulation of the node representations into the graph feature vector, the readout phase. In this process, all vector representations for all nodes are aggregated to make a graph-level feature representation. The final graph vector representation is defined as

$$h = \sum_{v \in G} \sigma(W_2 h_V^L) \tag{6}$$

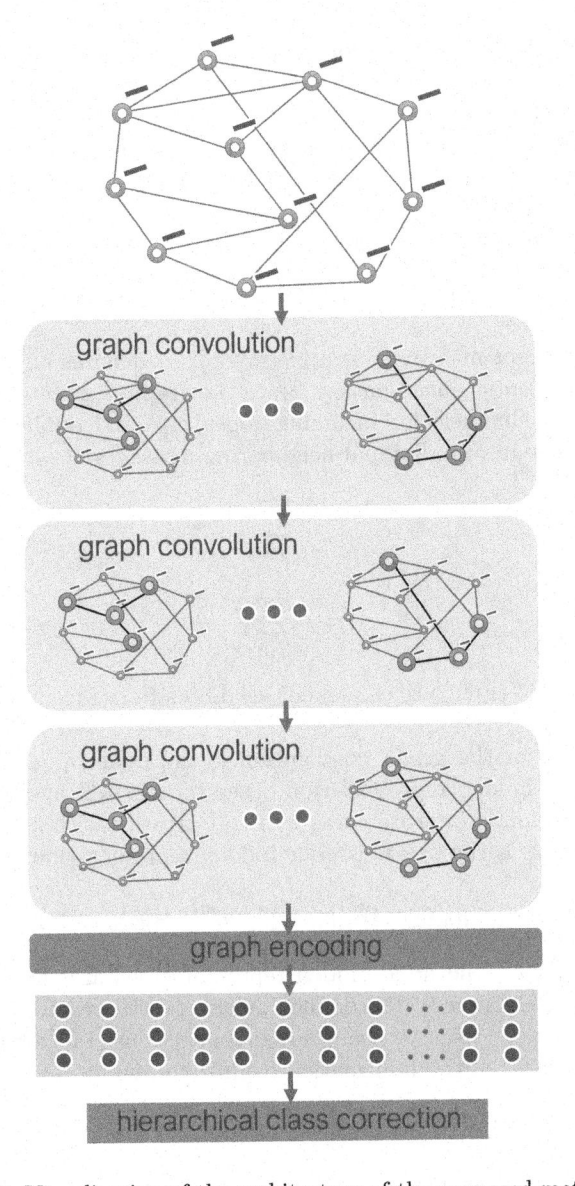

Fig. 2. Visualization of the architecture of the proposed method.

where h_v^L is the hidden representation of node v from the last layer. The node and graph representations are defined as 500-dimensional vectors.

The graph latent representation vector is input in the second part of the network, which consists of three fully connected (FC) layers of size 1024, 512 and N neurons, where N represents the number of terms (classes) in the corresponding ontology. The last layer gives the probabilities for each class as output, so a sigmoid function activates it. These settings create a standard multi-label classification, where each output is a value of 0 to 1.

The last layer is the Hierarchical Class Correction layer designed to capture the hierarchical multi-label classification property of the problem. This layer is composed of a 1-0 matrix with rows indicating the encoding of the relationship between the parent term and its descendant terms. That is, each term in the matrix has a vector where the elements at its position and the position of its descendants have value 1, and the remaining elements have value 0. Once the vector with predictions is element-wise multiplied with this matrix, there is an operation of max-pooling that extracts the maximum probability values for each term.

Benchmark Dataset. To create the train and test sets two timeframes of protein annotations are used: historical annotations t_0 from 05 July 2017 and current annotations t_1 from 13 February 2019. The approach in CAFA challenge [13] is applied to determine the separation of proteins into the training set and testing set. This method uses the proteins in both time annotations to create two kinds of sets for each GO ontology:

- No-knowledge data (NK): protein structures annotated with a term in current annotations t_1 and are not annotated in historical annotations t_0 with terms from an ontology;
- Limited-knowledge data (LK): proteins which are annotated with a term from an ontology in t_1 and are not annotated in the corresponding ontology in t_0.

The protein structures found in the NK and LK sets are part of the test set, that is, the benchmark set, while the remaining proteins are assigned to the train set. Table 1 shows the size of the train and test sets for every type of GO ontology.

Table 1. Number of annotated protein chain structures in the training and testing sets for every ontology.

Ontology	Train set	Test set	
		LK	NK
BP	66,495	231	13,797
		14,028	
MF	66,319	456	16,441
		16,897	
CC	59,060	137	13,408
		13,575	

The terms from each type of ontology that annotate less than ten protein chains are excluded from consideration in the training phase of the model. The model is not aware of these functional terms; accordingly, it is not able to learn to predict them.

Model Implementation and Optimization. The model was implemented using Tensorflow and Keras deep learning frameworks. The learning of the model was performed using the AMSGrad variant [19] of the Adam optimizer from [22] with a learning rate set to 0.001. The models are trained with 150 episodes with a batch size of 1, and 1000 steps per epoch. The most appropriate model is selected from the epoch with the lowest value of the loss function of the validation set, and accordingly, this version is used to evaluate the predictions of the particular model.

2.5   Evaluation

The evaluation of the protein-annotated method is based on the methods proposed in the Critical Assessment of the Protein Function Annotation Algorithms (CAFA) experiment [16,21]. Protein-based measures evaluate the results of classification methods that for a given protein predict the terms with which the corresponding protein has been annotated. Measures are needed that will assess the predictions of a model intended for multilabel classification. For this type of evaluation, the F_{max} measure, S_{min} measure, as well as the precision-recall curve, will be used.

The precision pr, the recall rc and F_{max} are calculated with the following formulas:

$$pr(t) = \frac{1}{m(t)} \sum_{i=1}^{m(t)} \frac{\sum_f I(f \in P_i(t) \wedge f \in T_i)}{\sum_f I(f \in P_i(t))} \tag{7}$$

$$rc(t) = \frac{1}{n} \sum_{i=1}^{n} \frac{\sum_f I(f \in P_i(t) \wedge f \in T_i)}{\sum_f I(f \in T_i)} \tag{8}$$

$$F_{max} = \max_t(\frac{2 \cdot pr(t) \cdot rc(t)}{pr(t) + rc(t)}) \tag{9}$$

where t is a decision threshold, $m(t)$ refers to the number of proteins for which at least one term is provided with a confidence greater than or equal to the defined threshold t, n is the number of proteins in the test set, $I(\cdot)$ is an indicator function, $P_i(t)$ is a set of predicted terms with a confidence greater than or equal to t for the protein i, while T_i is the set of real terms for the protein i. Mainly, the precision is the percentage of the predicted terms that are relevant, and the recall is the percentage of the relevant terms that have been predicted.

The other type of measures includes the remaining uncertainty ru, the misinformation mi, and the minimum semantic distance S_{min}. These measures are defined as:

$$ru(t) = \frac{1}{n} \sum_{i=1}^{n} \sum_f IC(f) \cdot I(f \notin P_i(t) \wedge f \in T_i) \tag{10}$$

$$mi(t) = \frac{1}{n} \sum_{i=1}^{n} \sum_f IC(f) \cdot I(f \in P_i(t) \wedge f \notin T_i) \tag{11}$$

$$S_{min} = \min_{t}(\sqrt{ri(t)^2 + mi(t)^2})$$ (12)

where $IC(f)$ is the informative content of the term f of the ontology.

3   Experiments: Discussion of Results

The main topic of this section is the discussion of the success of the proposed model and its comparison with related models designed to solve the same problem. The Naive method and the BLAST method are applied as baseline models to compare the proposed method in this project. The Naive approach predicts terms with their relative frequency in the training set; that is, each protein has the same predictions [8]. Consequently, if specific terms are more frequent in the train set, then these terms are predicted with higher confidence than other terms that are usually more explicit terms.

The BLAST method uses the Basic Local Alignment Search Tool – BLAST [2], which compares the protein sequences of the test set with the protein sequences of the training set. BLAST's basic idea is a heuristic search for alignment of sequences that have a high estimation, between the searched protein and the protein sequences from the database. The result after the search consists of the detected protein sequences along with the percentage of identical hits in the alignment. This value is used as a confidence value to annotate the protein with terms of the discovered proteins, that is, the terms are predicted with the corresponding value for identical hits. If one term is predicted multiple times (from various proteins), the highest confidence value is retained.

Table 2 contains the results of the experiments of protein function prediction with the test set. The results from this table are obtained by the same train and test sets defined previously, for training and testing respectively. For all ontologies, both the BLAST and Structure2Function methods are better than the random standard of the Naive approach.

The BLAST method achieved the highest F_{max} value for the BP ontology, although this value is close to the performance of Structure2Function model. However, according to the evaluation using the S_{min} metric, the Structure2Function method is a more refined method for this ontology. The S_{min} metric gives insight to the model that predicts more informative terms which are more desirable since it weights GO terms by conditional information content. Accordingly, the model with the lowest S_{min} value predicts more precise terms, than general terms.

The initial results of the Naive method for the ontology MF showed a high dominant F_{max} value, which is a consequence of a large number of proteins that are annotated only with the term GO: 0005515 (protein binding) and its parent GO: 0005488 (binding). This significant part of the exclusive annotations introduces great bias in the data, to the point where the Naive method has irrationally good results. Therefore, filtering the test set for the MF ontology has been made to remove proteins annotated only with these two terms. The train set applied to train the Structure2Function model was filtered in the same

Table 2. Results from the protein-based evaluation of the functional annotation methods with the test set.

Method	BP		MF		CC	
	Fmax	Smin	Fmax	Smin	Fmax	Smin
Naive	0.341	44.698	0.454	12.092	0.585	13.501
BLAST	**0.464**	49.555	0.445	**9.622**	0.646	16.397
Structure2Function	0.438	**39.871**	**0.492**	11.462	**0.653**	**12.010**

way, since the affinity of the set to these two terms was seen as a problem in the initial testing of the model. The evaluation for this ontology points the Structure2Function method as the best method according to the F_{max} metric, and S_{min} opposes the BLAST method as the best approach.

The CC ontology is the smallest ontology; namely, the proteins of the test set are annotated with a small number of relatively more general terms that are often used in the train set. For this ontology, the Structure2Function method is the superior approach according to F_{max} and S_{min} metric.

To visualize the results from Table 2, Fig. 3 depicts the curves for the ratio of precision-response measures. The perfect classifier would be characterized by $F_{max} = 1$ corresponding to the point $(1,1)$ in the precision-recall plane. These curves confirm the results and conclusions drawn from Table 2.

The train and test sets for protein annotation contain information on the annotations that have been confirmed by experiments, but there is a shortcoming in defining annotations that are not plausible, that is negative samples. Thus, in the predictions obtained with the models, new annotations of a given protein are received, and with the proposed evaluation measures they are assessed as wrong predictions, but there are no conditions to determine whether the new predicted terms represent a new knowledge or model error. Additional empirical support is required for new terms to eliminate this problem so that it is possible to determine with higher confidence whether these terms need to be rejected or confirmed.

The analysis of the experimental results demonstrated by the Structure2Function model proposed in this study validates the direction of further research. The GCN proves its ability to interpret protein secondary structure graphs, with amino acids as nodes and bonds as edges. Further work should include the improvement of this model as well as its extension. One drawback of this method is the limited number of proteins with known structures. The solution for this problem is predicting the secondary structure of the protein using its known amino acid sequence, which is already investigated with several existing methods. Therefore, our future work includes the problem of predicting the protein secondary structure using the primary structure, for the proteins with unknown secondary structure, and assigning functions to these proteins.

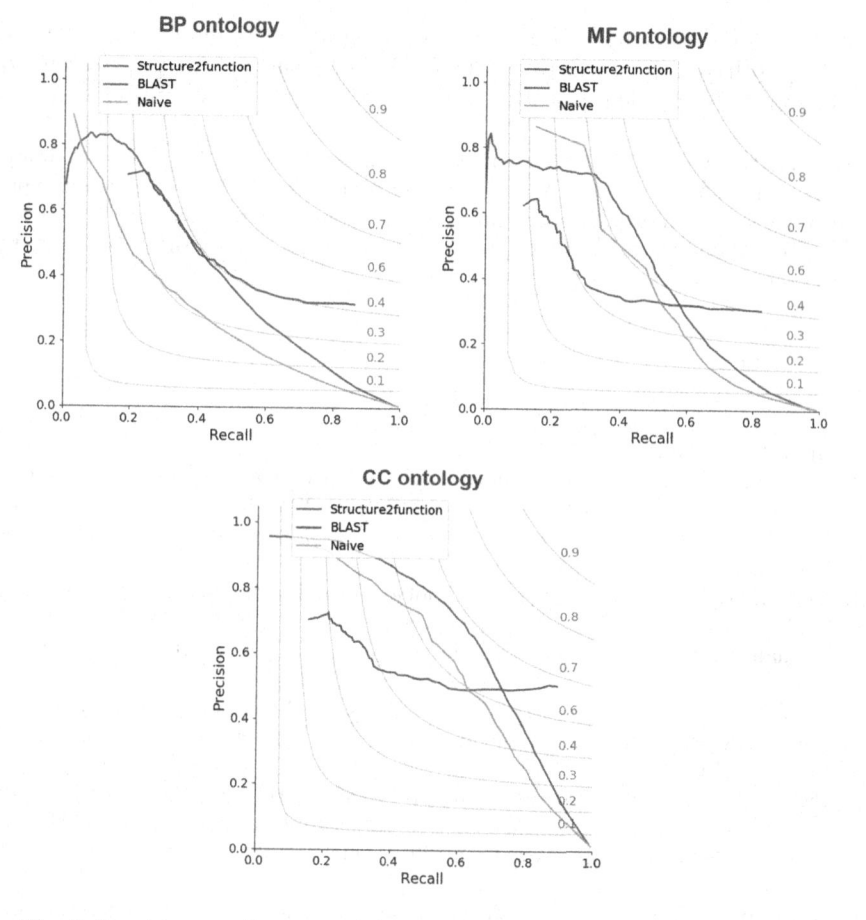

Fig. 3. Precision-recall curves of the experiments using the defined test set.

4   Conclusions

Proteins are vital components chargeable for carrying out functionalities in living organisms. That is why it is necessary to know their functions. Today, in contrast to a large number of known protein sequences, the number of known functional protein annotations are still in a small amount. Therefore, automatic detection of protein functions is an actively investigated area.

Recent research examines the protein function annotation with the aid of the information derived from protein sequences, protein interactions, and their structure. This study reviews the ability for protein annotation of a model consisted of Graph Convolutional Neural Networks (GANs), called Structure2Function. The model tries to map the patterns from protein secondary structure into protein function terms. The protein secondary structure is modeled in a novel way as a graph where the amino acids are represented as nodes, and the edges are the peptide and hydrogen bonds forming the secondary structure. The experi-

ments employed to compare the proposed method to baseline models verify the tested hypothesis of building an intelligent model proficient for protein function prediction, utilizing their secondary structure represented as a graph.

Acknowledgements. This work was partially financed by the Faculty of Computer Science and Engineering at the Ss. Cyril and Methodius University, Skopje, North Macedonia. The computational resources used for this research were kindly provided by MAGIX.AI and the NVIDIA Corporation (a donation of a Titan V GPU to Eftim Zdravevski).

References

1. Al-Lazikani, B., Jung, J., Xiang, Z., Honig, B.: Protein structure prediction. Curr. Opin. Chem. Biol. **5**(1), 51–56 (2001)
2. Altschul, S.F., et al.: Gapped blast and psi-blast: a new generation of protein database search programs. Nucleic Acids Res. **25**(17), 3389–3402 (1997)
3. Apweiler, R., et al.: Uniprot: the universal protein knowledgebase. Nucleic Acids Res. **32**, D115–D119 (2004)
4. Ashburner, M., et al.: Gene ontology: tool for the unification of biology. Nat. Genet. **25**(1), 25 (2000)
5. Berman, H.M., et al.: The protein data bank. Nucleic Acids Res. **28**(1), 235–242 (2000)
6. Bronstein, M.M., Bruna, J., LeCun, Y., Szlam, A., Vandergheynst, P.: Geometric deep learning: going beyond euclidean data. IEEE Signal Process. Mag. **34**(4), 18–42 (2017)
7. Chen, C., Huang, H., Wu, C.H.: Protein bioinformatics databases and resources. In: Wu, C.H., Arighi, C.N., Ross, K.E. (eds.) Protein Bioinformatics. MMB, vol. 1558, pp. 3–39. Springer, New York (2017). https://doi.org/10.1007/978-1-4939-6783-4_1
8. Clark, W.T., Radivojac, P.: Analysis of protein function and its prediction from amino acid sequence. Proteins Struct. Funct. Bioinf. **79**(7), 2086–2096 (2011)
9. Gene Ontology Consortium: The gene ontology (GO) database and informatics resource. Nucleic Acids Res. **32**(Suppl. 1), D258–D261 (2004)
10. Crooks, G.E., Brenner, S.E.: Protein secondary structure: entropy, correlations and prediction. Bioinformatics **20**(10), 1603–1611 (2004)
11. Dai, H., Dai, B., Song, L.: Discriminative embeddings of latent variable models for structured data. In: International Conference on Machine Learning, pp. 2702–2711 (2016)
12. Duvenaud, D.K., et al.: Convolutional networks on graphs for learning molecular fingerprints. In: Advances in Neural Information Processing Systems, pp. 2224–2232 (2015)
13. Gilmer, J., Schoenholz, S.S., Riley, P.F., Vinyals, O., Dahl, G.E.: Neural message passing for quantum chemistry. In: Proceedings of the 34th International Conference on Machine Learning, vol. 70, pp. 1263–1272. JMLR. org (2017)
14. He, K., Zhang, X., Ren, S., Sun, J.: Deep residual learning for image recognition. In: Proceedings of the IEEE Conference on Computer Vision and Pattern Recognition, pp. 770–778 (2016)
15. Jiang, Q., Jin, X., Lee, S.J., Yao, S.: Protein secondary structure prediction: a survey of the state of the art. J. Mol. Graph. Model. **76**, 379–402 (2017)

16. Jiang, Y., et al.: An expanded evaluation of protein function prediction methods shows an improvement in accuracy. Genome Biol. **17**(1), 184 (2016)
17. Kabsch, W., Sander, C.: Dictionary of protein secondary structure: pattern recognition of hydrogen-bonded and geometrical features. Biopolymers **22**(12), 2577–2637 (1983)
18. Kihara, D.: Protein Function Prediction: Methods and Protocols. Humana Press, Totowa (2017)
19. Kingma, D.P., Ba, J.: Adam: a method for stochastic optimization. CoRR abs/1412.6980 (2014)
20. Pearson, W.R.: Protein function prediction: problems and pitfalls. Curr. Protoc. Bioinform. **51**(1), 4–12 (2015)
21. Radivojac, P., et al.: A large-scale evaluation of computational protein function prediction. Nat. Methods **10**(3), 221 (2013)
22. Reddi, S.J., Kale, S., Kumar, S.: On the convergence of adam and beyond. In: Proceedings of the International Conference on Learning Representations (2018)
23. Rost, B.: Protein secondary structure prediction continues to rise. J. Struct. Biol. **134**(2–3), 204–218 (2001)
24. Yang, Y., et al.: Sixty-five years of the long march in protein secondary structure prediction: the final stretch? Briefings Bioinform. **19**(3), 482–494 (2016)
25. Zhang, M., Cui, Z., Neumann, M., Chen, Y.: An end-to-end deep learning architecture for graph classification. In: Thirty-Second AAAI Conference on Artificial Intelligence (2018)

Exploring the Attention Mechanism in Deep Models: A Case Study on Sentiment Analysis

Martina Toshevska[(✉)] and Slobodan Kalajdziski

Faculty of Computer Science and Engineering, Ss. Cyril and Methodius University, Skopje, Macedonia
{martina.toshevska,slobodan.kalajdziski}@finki.ukim.mk

Abstract. Interpreting what a deep learning model has learned is a challenging task. In this paper, we present a deep learning architecture relying upon an attention mechanism. The main focus is put on the exploratory evaluation of attention-based deep learning models on lexicons of affective words, and examination whether the word valence is the most significant information or not. Obtained evaluation results lead to a conclusion that word valences do play a significant role in sentiment analysis, but possibly models rely upon other dimensions perhaps not distinguishable by humans.

Keywords: Deep learning · Attention mechanism · Sentiment analysis · Sentiment lexicons · Word valence

1 Introduction

Analyzing people's attitude (opinion, sentiment, emotions) has received significant attention from the researchers. Variety of models for identifying sentiment (emotion, opinion) are already created with both machine and deep learning techniques.

Machine learning techniques rely mostly on feature engineering that sometimes depends on external resources. Lexicons of affective words and phrases are frequently used for extracting features which are later fed into a machine learning model [7,8,11]. Therefore, we can hypothesize that affective words and their associated information are an essential indicator of the sentiment. One type of information provided in lexical resources is valence. It refers to whether a word is considered positive or negative.

On the other side, deep learning models mostly rely only on the words present in the text [1,5,18]. Nevertheless, recent trends about incorporating lexical information into the deep model attract the researchers [16,17]. Deep learning models are built upon most common architectures recommended in the field of natural language processing - recurrent neural networks (RNNs) and convolutional neural networks (CNNs). Enhancing the model with an attention mechanism leads

© Springer Nature Switzerland AG 2019
S. Gievska and G. Madjarov (Eds.): ICT Innovations 2019, CCIS 1110, pp. 202–211, 2019.
https://doi.org/10.1007/978-3-030-33110-8_17

to promising performance improvement. The attention mechanism enables the model to focus on the part of the input.

However, what does the model, in fact, learn? Deep learning models achieve high accuracy at the expense of their interpretability since these models are often treated as black-box models. Analyzing deep learning models is often a challenging task. In what follows we highlight the findings of our experimental evaluation of various deep models. We examine their learned weights and to what extent they correlate to the information confirmed to provide reliable features for classical machine learning models. The rest of the paper is organized as follows. Section 2 describes in details the steps of our research. Discussion of experimental findings is provided in Sect. 3, while Sect. 4 concludes the paper.

2   Research Methodology

A broad research of understanding and visualizing the knowledge of a model already exists in many domains. Depending of the context, different part of the model is being analysed, starting from visualizing the output of interme-diate layers in generative adversarial networks [4] to interpreting convolutional and recurrent neural networks for natural language processing [2]. Of primary importance in this study is to explore the attention mechanisms [3] in the domain of sentiment analysis (sentiment classification and star detection). According to [9] attention weights should correlate with feature importance measures. In sen-timent analysis, as previously stated, lexical information is proven to provide reliable features. We hypothesize that such information correlates with learned attention weights.

Following the techniques provided by [9], we have conducted several exploratory studies to examine whether attention weights correlate to word valences provided by lexical resources. For that purpose, we utilize a dataset primed for sentiment analysis and various lexicons providing word valence. We develop several deep learning models based on an attention mechanism. In the end, we analyze the weights of the trained models. Our key research questions are:

1. Does the attention layer learn the importance of information encoded in words such as valence?, and
2. To what extent do word valences correlate with attention weights?

2.1   Dataset

In this study, we utilize the Yelp[1]dataset. This dataset consists of more than 6M reviews obtained from Yelp. We apply several techniques for filtering and preprocessing the dataset. The following subsections describe in detail these techniques.

---

[1] https://www.yelp.com/dataset/challenge, last accesed: 13.06.2019.

Dataset Filtering. The dataset is composed of textual reviews each annotated with the ID of the user that wrote it, the business id it is written for, star rating, date and number of received votes. The total number of distinct users is approximately 1.5M. Figure 1 plots the distribution of reviews per user on logarithmic scale. The number of reviews per user fluctuates, starting with users that wrote 1 review up to users that wrote 4,129 reviews. Assuming that some users write too many reviews that are often fake, we eliminate reviews from users with a high frequency of written reviews such that only reviews written by users with review frequency in ranges [50, 500] and [10, 500] remain. For clarity, these subsets are denoted as Subset A and Subset B, respectively.

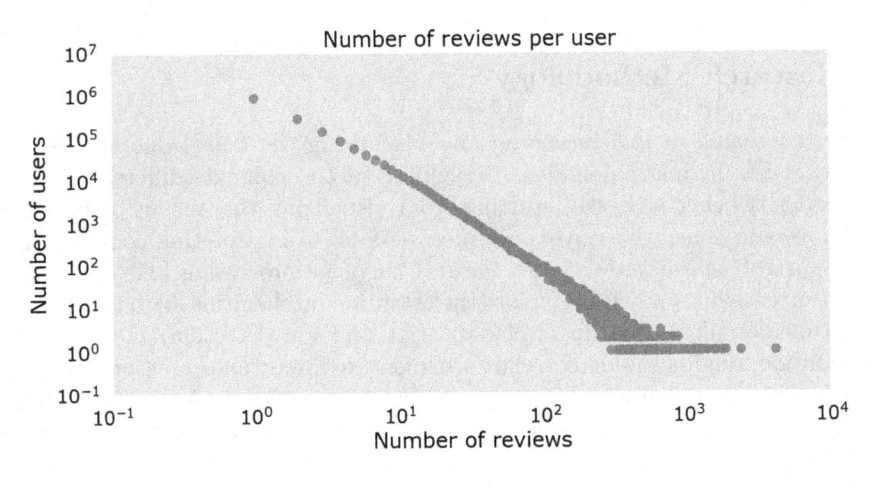

Fig. 1. Number of reviews per user.

Furthermore, we apply filtering by review length. The maximum length is 1,870, while the minimum is 1. The distribution of review length is plotted on logarithmic scale and is shown in Fig. 2. We keep only reviews with length in the range from the 25th percentile to 75th percentile, that is [50, 162]. The total number of reviews after filtering is summarized in Table 1.

Table 1. Number of reviews remaining after filtering.

Stars	All	Subset A (50–500)	Subset B (10–500)
1	1,002,159	39,984	165,334
2	542,394	52,255	140,111
3	739,280	116,529	238,894
4	1,468,985	224,652	475,143
5	2,933,082	216,340	678,278

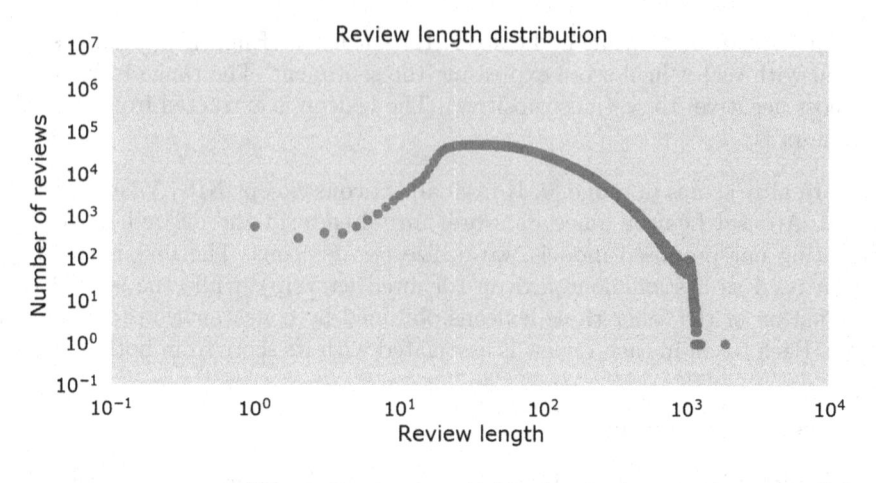

Fig. 2. Review length distribution.

Data Pre-processing. Reviews consist of multiple sentences. Since our analysis is based on words and not sentences, we consider each review as one sentence. The first step is applying necessary pre-processing steps: lower-casing and tokenizing each review. Subsequently, we extract the base form of words by applying part-of-speech tagging and lemmatization. The final pre-processing step is filtering the vocabulary. We filter the vocabulary in the following way:

- Remove words occurring in less than 15 reviews
- Remove punctuation
- Remove stopwords

The filtering is done by merely deleting tokens that satisfy the filtering criteria leading to a reduction of the number of words in a review. For Subset A, the average review length drops from 100 to 46, and from 96 to 44 for Subset B.

Sentiment Lexicons. Our analysis relies on several lexical resources i.e. lexicons.

- AFINN[2] - a list of 2,477 English words and phrases annotated with their valence rating, an integer value between -5 and 5 denoting the strength of the emotion expressed with a word.
- NRC Valence, Arousal, Dominance Lexicon [14] - a list of 20,000 English terms associated with valence, dominance, and arousal score. The score is between 0 and 1. From this lexicon, we exploit only the valence scores.
- NRC Hashtag Sentiment Lexicon [12] - a list of English n-grams annotated with real-valued score expressing the sentiment. The range is from $-\infty$ (most negative) to ∞ (most positive). The lexicon is extracted from tweets with sentiment word hashtags.

---

[2] http://corpustext.com/reference/sentiment_afinn.html, last accesed: 18.08.2019.

- Yelp Restaurant Sentiment Lexicon [12] - a list of English n-grams associated with real-valued score expressing the sentiment. The range is from $-\infty$ (most negative) to ∞ (most positive). The lexicon is extracted from the Yelp dataset.

We normalize scores in range [0, 1] from all lexicons except NRC Valence, Dominance, Arousal Lexicon since its scores are already in the desired range. For evaluating our proposed models, we utilize two lexicons. The first is the Yelp lexicon used as a standalone lexicon (denoted as Yelp), while the second is a combination of the other three lexicons obtained by concatenation (denoted as NRC). Each token in each review is associated with its score from both lexicons.

2.2 Deep Architectures

This section explains in detail the architectures of our proposed models. The overall architecture for all models is shown in Fig. 3. The input is a sequence of tokens composing each review. The first layer is an Embedding layer for constructing matrix representation of the input sequence. Its dimension is $m \times n$ where m is the number of tokens in the input sequence and n is the word embedding size. For the number of tokens, we use the median of the review length in the corresponding data subset, namely 45 for Subset A and 42 for Subset B.

Word embeddings represent each word in the vocabulary as a dense real-valued vector, while preserving word meanings, their relationships, and the semantic information. Pre-trained word embedding vectors already exist. We use GloVe (Global Vectors for Word Representation) [15] embedding vectors. In this study, we utilize two different word embeddings as initializers for the Embedding layer: 300-dimensional vectors pre-trained on Wikipedia and 200-dimensional vectors pre-trained on Twitter.

The matrix representation created with the Embedding layer is fed into a recurrent layer. As a recurrent layer, for all models, we use Long Short-Term Memory networks (LSTMs) with 1,024 units. The next layer is an attention mechanism inspired by [6,13]. It performs attention over hidden states of the recurrent layer. The attention outputs feature vector of attention weights for each input token which are multiplied with token representation obtained from the LSTM.

The final part of the models is Multilayer Perceptron encompassing four fully connected layers, each followed by a Dropout layer with rate 0.5. The first two layers are composed of 1,024 units, while the third and the fourth are composed of 512 and 256 units, respectively. The output is a class for the review. We train models with two different versions for the class. The first version is classifying the review as either positive or negative[3] (the model is denoted as SentDetect). The second version is a model denoted as StarDetect, which determines the number of stars that the review receives.

---

[3] A review is considered positive if it received at least 4 stars, and negative otherwise.

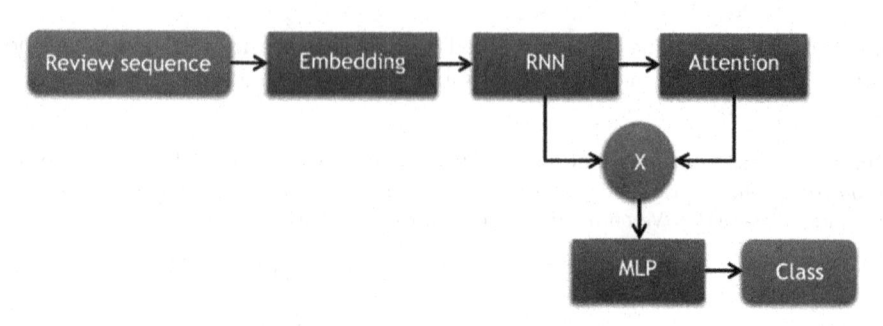

Fig. 3. Overall architecture for the models.

All models were trained with categorical cross-entropy loss function, and Adam optimizer [10] with 0.0001 learning rate and batch size 256. The training was performed on NVIDIA GeForce GTX TITAN X in 50 epochs.

3 Experiments and Discussion of Results

To test our hypothesis, we run a battery of experiments. Each subset is partitioned into training, validation and test sets in a ratio of 70:15:15. This section presents evaluation results of our models. We report on two evaluation metrics: correlation coefficient (Pearson and Kendall), and Jaccard similarity. Evaluation results presented in the following subsections are obtained on test sets.

3.1 Correlation Between Attention Weights and Word Valences

We measure the extent to which word valences correlate with attention weights by calculating correlation coefficient. The output of the attention mechanism assembles an attention matrix composed of attention weights for each word at each timestep. Single attention weight for a particular word is obtained by summing all weights for that word, creating one-dimensional attention vector for each review. With word valences we create one-dimensional valence vector with values from the NRC lexicon and one-dimensional valence vector with values from the Yelp lexicon. All three vectors have the same length that is the minimum of the number of words in the review and the padding size. In fact, we do not take into account words that have not contributed to the final prediction when the length is greater than or equal to the padding size, and omit padded zeros when the length is less than the padding size.

Table 2 summarizes our findings about SentDetect and StarDetect model, respectively. We report on two correlation metrics - Pearson correlation coefficient and Kendall's tau coefficient. Correlation coefficients are calculated between attention vector and: (1) NRC valence vector and (2) Yelp valence vector for each review in the set. The final coefficient is computed by averaging

over the obtained values. The observed correlations are near 0^4 indicating no correspondence between word valences and attention weights.

Table 2. Correlation coefficient for SentDetect and StarDetect models. Pearson correlation coefficient - r, Kendall correlation coefficient - τ. GloVe embeddings pre-trained on Wikipedia - w, GloVe embeddings pre-trained on Twitter - t.

			NRC		Yelp	
			r	τ	r	τ
SentDetect	Subset A	w	−0.00646	−0.00482	−0.00026	−0.00201
		t	0.0246	0.02369	0.02872	0.02428
	Subset B	w	−0.01277	−0.00633	−0.00368	−0.00185
		t	−0.00899	−0.00957	−0.00072	−0.00193
StarDetect	Subset A	w	−0.02745	−0.01291	−0.03327	−0.01976
		t	0.00124	−0.00632	0.04243	0.03106
	Subset B	w	−0.00043	0.00332	0.00473	0.00261
		t	−0.00078	0.00123	0.00203	0.0022

3.2 Jaccard Similarity

We further investigate whether words with high attention weights are those having high lexical valences. For this purpose, words are sorted by two criteria: attention weight and valence. The first criteria simply assumes that higher weight value implies high attention weight. For the second criteria we need to consider two different aspects, whether the review is positive or negative. We assume that higher lexical value implies higher valence if the review is positive or has more than three stars, while lower lexical value implies higher valence if the review is negative or has less than four stars.

After that, we compute the overlap of the set of words with highest attention weights and the set of words with highest valences. The overlap is defined as Jaccard similarity between the two sets of words. We have no clear definition of what highest weights and highest valences stand for i.e. how many words are those with highest weights and how many words are those with highest valences. Therefore, we report Jaccard similarity at a specific cutoff. For instance, if the cutoff is set to 5, Jaccard similarity between the set of first 5 words with highest attention weights and the set of first 5 words with highest lexical valences is computed. Note here that computing Jaccard similarity at cutoff equal to padding size (45 for Subset A and 42 for Subset B) leads to value 1 for similarity since, although having different ordering, both sets of words are equal.

The Jaccard similarity is calculated at four different cutoff sizes: 5, 10, 15 and 20. Table 3 reports summary statistics for evaluation on the NRC lexicon,

---

[4] Value 0 implies that the two sets are not correlated.

while Table 4 reports summary statistics for evaluation on the Yelp lexicon. For the first cutoff size, 5, the similarity is approximately 0.08 which indicates that the average number of mutual words is less than 1. Values increase as the cutoff size increases which is reasonable since the sets are expanding too. We could hypothesize that both sets will have more mutual words by taking into account more words. For example, the similarity at cutoff 20 is approximately 0.4 indicating that the average number of mutual words is around 8. According to the findings, we can conclude that word valence, despite playing significant role in sentiment analysis, is not the most significant information in attention-based deep learning models.

Table 3. Jaccard similarity for SentDetect and StarDetect models evaluated on the NRC lexicon. GloVe embeddings pre-trained on Wikipedia - w, GloVe embeddings pre-trained on Twitter - t.

			NRC			
			First 5	First 10	First 15	First 20
SentDetect	Subset A	w	0.0803	0.1678	0.27335	0.40283
		t	0.0802	0.1676	0.27421	0.4039
	Subset B	w	0.08722	0.17823	0.28883	0.42577
		t	0.0891	0.17649	0.28783	0.42621
StarDetect	Subset A	w	0.07991	0.16722	0.27439	0.40549
		t	0.08593	0.17326	0.27772	0.40707
	Subset B	w	0.08604	0.17669	0.28695	0.42336
		t	0.08533	0.17594	0.28636	0.42271

Table 4. Jaccard similarity for SentDetect and SentDetect models evaluated on the Yelp lexicon. GloVe embeddings pre-trained on Wikipedia - w, GloVe embeddings pre-trained on Twitter - t.

			Yelp			
			First 5	First 10	First 15	First 20
SentDetect	Subset A	w	0.08295	0.1686	0.27308	0.4021
		t	0.083	0.16938	0.27508	0.40476
	Subset B	w	0.08932	0.179	0.29013	0.42665
		t	0.08855	0.17744	0.28812	0.42561
StarDetect	Subset A	w	0.08249	0.16777	0.27374	0.40507
		t	0.09034	0.17525	0.27799	0.40675
	Subset B	w	0.08715	0.17774	0.28723	0.42331
		t	0.08719	0.17702	0.28676	0.42342

4   Conclusion

In this paper, we aim to explore the attention mechanishm and investigate its learned weights. We proposed a deep learning architecture based on attention mechanism and trained different models on two sentiment analysis tasks: review sentiment classification and review star detection.

Information about affective words and phrases, and their assigned valence scores provide reliable features for machine learning models. Moreover, when incorporated into deep learning models, such lexical information improves the performance of the models. The key task in this paper is to examine whether the attention mechanism is able to capture the importance of word valences and whether the attention weights correlate to word valences provided by lexical resources.

By performing various experiments, the findings suggest that word valence is not the most significant information in attention-based deep learning models. Obtained evaluation results lead to conclusion that word valences do play significant role in sentiment analysis, but possibly models rely upon other dimensions perhaps not distinguishable by humans.

References

1. Abdul-Mageed, M., Ungar, L.: Emonet: fine-grained emotion detection with gated recurrent neural networks. In: Proceedings of the 55th Annual Meeting of the Association for Computational Linguistics (vol. 1: Long Papers), pp. 718–728 (2017)
2. Alishahi, A., Chrupala, G., Linzen, T.: Analyzing and interpreting neural networks for NLP: a report on the first blackboxnlp workshop. Nat. Lang. Eng. (2019)
3. Bahdanau, D., Cho, K., Bengio, Y.: Neural machine translation by jointly learning to align and translate (2014)
4. Bau, D., et al.: GAN dissection: visualizing and understanding generative adversarial networks (2018)
5. Chen, H., Sun, M., Tu, C., Lin, Y., Liu, Z.: Neural sentiment classification with user and product attention. In: Proceedings of the 2016 Conference on Empirical Methods in Natural Language Processing, pp. 1650–1659 (2016)
6. Giannakopoulos, A., Antognini, D., Musat, C., Hossmann, A., Baeriswyl, M.: Dataset construction via attention for aspect term extraction with distant supervision. In: 2017 IEEE International Conference on Data Mining Workshops (ICDMW), pp. 373–380. IEEE (2017)
7. Gievska, S., Koroveshovski, K., Chavdarova, T.: A hybrid approach for emotion detection in support of affective interaction. In: 2014 IEEE International Conference on Data Mining Workshop, pp. 352–359. IEEE (2014)
8. Hasan, M., Rundensteiner, E., Agu, E.: Emotex: detecting emotions in Twitter messages. In: Proceedings of the Sixth ASE International Conference on Social Computing (SocialCom 2014). Academy of Science and Engineering (ASE) (2014)
9. Jain, S., C. Wallace, B.: Attention is not explanation. CoRR (2019)
10. Kingma, D.P., Ba, J.: Adam: a method for stochastic optimization. CoRR abs/1412.6980 (2014)

11. Kiritchenko, S., Zhu, X., Cherry, C., Mohammad, S.: NRC-Canada-2014: detecting aspects and sentiment in customer reviews. In: Proceedings of the 8th International Workshop on Semantic Evaluation (SemEval 2014), pp. 437–442 (2014)
12. Kiritchenko, S., Zhu, X., Mohammad, S.M.: Sentiment analysis of short informal texts. J. Artif. Intell. Res. **50**, 723–762 (2014)
13. Lin, Z., et al.: A structured self-attentive sentence embedding. CoRR (2017)
14. Mohammad, S.: Obtaining reliable human ratings of valence, arousal, and dominance for 20,000 English words. In: Proceedings of the 56th Annual Meeting of the Association for Computational Linguistics (vol. 1: Long Papers), pp. 174–184 (2018)
15. Pennington, J., Socher, R., Manning, C.: Glove: global vectors for word representation. In: Proceedings of the 2014 Conference on Empirical Methods in Natural Language Processing (EMNLP), pp. 1532–1543 (2014)
16. Shin, B., Lee, T., Choi, J.D.: Lexicon integrated CNN models with attention for sentiment analysis. In: Proceedings of the 8th Workshop on Computational Approaches to Subjectivity, Sentiment and Social Media Analysis, pp. 149–158 (2017)
17. Stojanovska, F., Toshevska, M., Gievska, S.: Explorations into deep neural models for emotion recognition. In: Kalajdziski, S., Ackovska, N. (eds.) ICT 2018. CCIS, vol. 940, pp. 217–232. Springer, Cham (2018). https://doi.org/10.1007/978-3-030-00825-3_19
18. Tang, D., Qin, B., Liu, T.: Document modeling with gated recurrent neural network for sentiment classification. In: Proceedings of the 2015 Conference on Empirical Methods in Natural Language Processing, pp. 1422–1432 (2015)

Image Augmentation with Neural Style Transfer

Borijan Georgievski[1,2]([⊠]) [iD]

[1] Netcetera, Skopje, North Macedonia
borijangeorgievski@protonmail.com
[2] FCSE, Ss. Cyril and Methodius University, Skopje, North Macedonia

Abstract. The amount of training data is of crucial importance for the per-formance of machine learning, and especially deep learning models. It is one of the most important factors that determine whether the developed model is effective or not. When the quantity of training data for a computer vision problem is insufficient, various data augmentation techniques are used to arti-ficially extend the training dataset with samples that retain the natural distri-bution of the original data. This paper proposes and evaluates a deep learning model that will be used for image augmentation. A complex deep neural net-work makes use of transfer learning in order to learn the characteristics of the content and style of the training images, create random style embeddings via learned multivariate normal distribution, and ultimately generate images to extend the original dataset. The model is trained on two datasets which are frequently used in computer vision: ImageNet and Painter by Numbers (PBN). Afterwards, the model is used to generate new images from the CIFAR-100 and Tiny-ImageNet-200 datasets. The performance of the augmentation model is evaluated by a separate convolutional neural network. The evaluation model is trained on the combined dataset, consisting of both, the original and augmented images, and then compared to the performance of the same model trained on the original datasets.

Keywords: Image augmentation · Neural style transfer · Computer Vision Convolutional Neural Networks (CNNs) · Deep learning

1 Introduction

In the current era of deep learning, there is one thing that can always improve a developed model, and that is more data. On one hand, deep learning models are becoming more accurate than every other carefully developed and hand-designed machine learning method, but on the other hand, they also need much more data. Having a small dataset is one of the biggest setbacks in computer vision projects, especially those with a deep learning approach. This paper focuses on exploring image manipulation through neural style transfer, and adopts an approach for randomizing style, in order to achieve arbitrary image augmentation.

Neural style transfer is a technique for reconstructing images by changing their style. It all started when Gatys et al. (2015) showed the possibility of using convolu-tional neural networks for transforming images, such that they are altered by applying

S. Gievska and G. Madjarov (Eds.): ICT Innovations 2019, CCIS 1110, pp. 212–224, 2019.
https://doi.org/10.1007/978-3-030-33110-8_18

styles of other chosen images, whilst preserving their content [2]. A steady progress has been made in a number of research studies [1, 3, 5, 15] since the original idea was proposed. The current state-of-the-art models can generate a new image based on content and style input images in a single forward pass. Even though, neural style transfer was intended for creating appealing images by combining content and style images of our choosing, the ability to extract style and content separately from an image seems to have promising uses in image augmentation. Although the pioneering approaches were usually limited in the number of styles that can be applied to images, nowadays the content and style images can be completely arbitrary. Stylizing an image requires us to have an already sampled style image, but a data augmenter capable of generating only a limited (or finite) number of different augmentations may be undesirable.

The model presented in this paper is not limited to any number of styles. Arbitrary augmentation is achieved by using a style embedding as style input instead of a real image. In this manner, the style embedding can still be extracted from a style image, or like in our case, it can be randomized. A pretrained neural network is fine-tuned and trained to extract a style embedding from an image, so that afterwards the channel-wise mean and covariance matrix of the styles observed in training phase are used to define a multivariate normal distribution. Therefore, this distribution can be sampled for an arbitrary number of times to generate a randomized style embedding which still maintains the natural distribution of the style images dataset. The complex loss function defined for this model uses a pretrained VGG19 model to extract the content and styles from the input and output images. Finally, the model's performance is evaluated with a separate convolutional neural network on the CIFAR-100 and Tiny-ImageNet-200 datasets.

2   Related Work

This section provides theoretical analysis of the concepts and methods used in this paper. A summary of previous related research on data augmentation and neural style transfer will provide a background against which this work should be positioned and compared.

2.1   ImageNet and Successful Architectures

Nowadays, there are many excellent neural networks which are available to the public, either as pretrained networks, or as architectures that only require data for the input, so that the users can train them themselves. Usually, the pretrained models are benchmarked at the annual ImageNet Large Scale Visual Recognition Challenge (ILSVRC) competition, active since 2010. Each year, the ImageNet organizers release high-quality, substantial dataset to be used for training, and many researchers use this data for other projects. The ILSVRC2017 image dataset contains over 200 GB of images, having two separate datasets for object localization and object detection. The two datasets combined contain approximately two million images. Our style transformer module is being trained with 537K images from this dataset.

The models which achieve best results at the annual competition are very often released to the public as pretrained networks, making their utilization very convenient to the independent researchers. One such successful model available for use is VGG [12], which is utilized in the proposed model as a pretrained loss network to extract low-level and high-level semantic representations. The input size of the model is 224 × 224 on three color channels (RGB), and only 3 × 3 convolutions and 2 × 2 pooling are used throughout the whole network. Every convolution in the VGG architecture uses rectified linear activation (ReLU). VGG also shows that the depth of the neural network plays an important role, in the sense that deeper networks give better results. One drawback of VGG is that this network is very big and resource heavy, containing around 160 million trainable parameters. In our model, the 19-layer pretrained VGG network is used as a feature extractor for the loss function.

He et al. (2015) find that training extremely deep neural networks are very hard to train because of vanishing and exploding gradient problems, but also propose a method for allowing to make extremely deep convolution neural networks (up to 152 layers) trainable [4]. The residual networks described in the paper yielded the best results at many subtasks of the ILSVRC 2015 competition (by the team name MSRA). Their main groundbreaking idea, which is also used in this paper, is the introduction of the residual block. It is a building block that can be implemented in very deep CNNs, where it can perform an identity mapping from a shortcut (or a skip connection), adding the shortcut to a feature map that is found several layers deeper, resulting in preserving past information in networks with even 100+ layers.

Szegedy et al. (2016) release the fourth iteration of Inception (Inception-V4) and Inception-ResNet-V2 which combines the best out of the two models, by utilizing the residual blocks inside the Inception architecture [13]. This model achieves superb Top5 accuracy of 95.3 on the ILSVRC-2012-CLS dataset, while VGG16 (16-layer model) and VGG19 (19-layer model) achieve 89.8 Top-5 accuracy. Kornblith et al. (2018) show that when the networks are used as fixed feature extractors or when they are fine-tuned, there is a strong correlation between the ImageNet accuracy and transfer accuracy, r = 0.99 and 0.96, respectively, suggesting that better ImageNet architectures are capable of learning better, transferable representations [9]. That is why the InceptionResNet-V2, pretrained on the ImageNet dataset, is used as a feature extractor for the style predictor module.

2.2 Neural Style Transfer

One of the most exciting applications of CNNs in the last years has been neural style transfer. Neural style transfer is a technique that extracts the content from an arbitrary image (content input), extracts the style of another image (style input), and generates a new image by combining the extracted content and style. Gatys et al. derive the neural representations of the content and style of an image from the feature responses of a 19-layer VGG network trained on object recognition. The feature space is provided by 16 convolutional layers and 5 average-pooling layers. The actual learning is performed by first initializing the generated image G randomly, and then iteratively improving the image by using gradient descent to minimize the cost function. They have designed the network such that the generated image G is an input to the network, modifying it after

each forward propagation based on the total loss. The overall optimization objective is defined as a combination of the style loss, and content loss [2]:

$$\mathcal{L}_{total} = \alpha\mathcal{L}_c(x, c) + \beta\mathcal{L}_s(x, s) \tag{1}$$

where $\mathcal{L}_c(x, c)$ and $\mathcal{L}_s(x, c)$ are the content loss and style loss components, while α and β are parameters which define the weights of the style and content losses. These two loss components use a pre-trained CNN (VGG19), in order to extract semantic features from the images.

Although this pioneering approach yielded good results, it is very computationally inefficient. In order to generate a single image given a content and style input, this model would need multiple forward propagations before it converges. To generate a second image, all this needs to be repeated once again. However, since the release of this paper there are many alternate approaches, many of which succeed in generating an image through a single forward pass.

Fast approximations with feed-forward neural networks have been proposed to speed up neural style transfer. Yanai (2017) proposes a model which generates a stylized image through a single forward pass. This approach involves a style condition network which generates a conditional signal from a style image directly. By adding a CNN that takes a style image as an input and outputs a conditional signal, the whole network can learn unlimited number of styles. To achieve style transfer for unseen styles, the CNN is expected to generate a conditional style signal by combining the conditional signals of the trained styles [15].

Ioffe et al. (2015) show how including batch normalization or instance normalization as part of the model architecture allows us to use much higher learning rates and be less careful about initialization, by performing normalization for each training mini-batch [6]. Ulyanov et al. (2017) show that by using instance normalization instead of regular batch normalization, it is possible to dramatically improve the performance of deep neural networks for image generation [14].

Huang et al. (2017) present an arbitrary style transfer approach for stylization in real-time by introducing an adaptive instance normalization (AdaIN) layer that adjust the mean and variance of the content features to match those of the style features. Most importantly in our case, these normalization techniques also act as a regularizer. Similarly to regular instance normalization, the mean μ and variance σ are computed across the spatial axes of an encoder network applied to a style image. However, unlike batch normalization, instance normalization and conditional instance normalization (detailed below), the AdaIN layer has no learnable affine parameters. Instead, it adaptively computes the affine parameters from the style input [5]:

$$AdaIN(x, y) = \sigma(y)\left(\frac{x - \mu(x)}{\sigma(x)}\right) + \mu(y) \tag{2}$$

Following up on instance normalization, Dumoulin et al. (2017) propose a conditional instance normalization, which generates a normalized activation $CIN(x, s)$, such that it learns a different set of parameters γ^s and β^s for each style s [1]:

$$CIN(x, s) = \gamma^s \left(\frac{x - \mu(x)}{\sigma(x)} \right) + \beta^s \tag{3}$$

where μ and σ are the mean and variance of the input x across the spatial axes. With the use of conditional instance normalization, Ghiasi et al. (2017) propose a method for fast and arbitrary style transfer in real-time. They build a style transfer network $S(\cdot)$ as an encoder-decoder, such that it shares its representation across many paintings, providing a rich vocabulary for representing any painting. Next, a style prediction network $P(\cdot)$ predicts an embedding vector S from an input style image, which supplies a set of normalization constants for the style transfer network. The advantage of this approach is that the model can generalize to an unseen style image by predicting its proper style embedding at test time [3]. They employ a pretrained Inception-v3 architecture and compute the mean across each activation channel of the truncated model's output which returns a feature vector with the dimension of 768. Afterwards, they apply two fully connected layers on top of the pretrained network to predict the final embedding S.

Following the approach of Dumoulin et al., Ghiasi et al. employ conditional instance normalization to normalize activation channels using the embedding vector S. However, this method learns the mapping from the style image to style parameters directly, as opposed to providing a fixed heuristic mapping from style image to normalization parameters. The content and style losses are derived from the distance in representational space of the VGG image classification network. Their model architecture is used in this paper, and it was selected from the alternative approaches mainly because the model is very intriguing and intuitive. In addition, the paper provides detailed information for the model configuration.

2.3 Image Augmentation

The generalizability of any machine learning algorithm is very often determined by the size of the training datasets, the more training data are used the better generalization is expected from the model. Unfortunately, in practice, the amount of data available for a project is limited. One way to solve this problem is by increasing the size of the training set by adding artificially or synthetically created data to the training set.

Models can benefit from simple augmentation techniques, such as: random translations, random rotations, horizontal/vertical flips, scaling and blurring. Simple operations like these can often greatly improve generalization, even if the model has already been designed to be translation invariant by using the convolution and pooling techniques that were discussed in previous sections/chapters. More advanced transformations exist, such as, nonlinear geometric distortions of the input or altering the intensities of the RGB channels in training images along with image translations and horizontal reflections [10].

Adding noise in the input to a neural network is also a form of data augmentation. For many classification and regression tasks, the task should still be possible to solve, even if small random noise is added to the input. Noise injection can also work when the noise is applied to the hidden units in a network, which can be interpreted as

performing a dataset augmentation at multiple levels of abstraction. Dropout can also be seen as a process of constructing new inputs through multiplying by noise. Data augmentation is effective for speech tasks as well, where the speech can be synthesized with noise in order to mimic actual human speech [11].

Jackson et al. (2018) explore the option of augmenting images using neural style transfer. They adapt the style transfer model architecture proposed by Ghiasi et al. to perform style randomization, by sampling input style embeddings from a multivariate normal distribution instead of inferring them from a style image [8]. During training, the style augmentation randomizes texture, contrast, and color, while preserving shape and semantic content. They chose the approach of Ghiasi et al., for its speed, flexibility, and visually compelling results. This research is of crucial importance for our work, as the model architecture very closely follows the ideas described in the paper.

3 Methodology

This section is dedicated to demonstrating the construction of the model, the learning phase and the project setup. The model progress throughout the training epochs will also be presented.

3.1 Model

The model structure, which is based on the approach proposed by Ghiasi et al., can be presented as a form of complex concatenation of two networks. One of them is the style predictor (·), which has the task of predicting the style embedding vector S, based on the input image. The style predictor uses the Inception-ResNet-V2 model, as a feature extractor, pretrained on the ImageNet dataset. The spatial hierarchy of features learned by the pretrained network effectively acts as a generic model of the visual world. This model is truncated at the layer mixed-6a, the output of which takes the shape of (Batch size, 17, 17, 1088), where we have 1088 features for each point in the 17×17 spatial map. The output of the layer mixed-6a is mean pooled across the tensor's spatial axes. The resulting tensor is finally passed through a 1×1 convolutional layer that produces a style embedding of shape (Batch size, 1, 1, 100). The low-level layers of this pretrained network would capture very low-level (and thus not useful) features, while the high-level layers would capture very complex concepts. The downside from extracting features deep in the network is that they are getting increasingly specific to the task that the model was previously trained on.

The style predictor has a total of 4,451,140 parameters, but only 108,900 of them are trainable. Most of the features are frozen, which is good for maintaining the complexity of this network as small as possible, considering the style transformer and the loss network (VGG19) are computationally expensive to run. The frozen layers do not update any parameters during backpropagation, and this is useful for many different reasons, such as: reducing training time by having less trainable parameters [7] and for achieving better results since the pretrained model was trained on a much bigger dataset.

The second network is called style transformer, which is a CNN in the form of encoder-decoder that uses residual blocks. It receives as input an RGB image, which is an array of shape (256, 256, 3), and produces a stylized RGB image of the same shape. The input is preprocessed such that the pixels are scaled between −1 and 1, sample wise. Furthermore, the input images use reflection padding instead of zero-padding, in order to eliminate border artifacts. The style embedding vector S, generated from the style predictor, controls the style transformer via conditional instance normalization (Eq. 3). The values of γ^s and β^s are calculated by passing S, once for each parameter, through a 1×1 convolutional layer, which acts like a memory efficient fully connected layer.

In contrast to the model presented in Ghiasi et al., in which a style image is taken as an input for each content image, S is a weighted combination between a random embedding and the style of the input (content) image. This allows us to generate an arbitrary number of styles during run-time, ultimately augmenting the data. The random embedding vector is sampled from a multivariate normal distribution with defined mean and covariance matrix that are calculated as the channel-wise metrics for all training style images, which were observed during the style predictor's training phase. Therefore, the final embedding is a function of the content image c [8]:

$$S = \alpha \mathcal{N}\left(\sigma, \sum\right) + (1 - \alpha)P(c) \tag{4}$$

P(c) is the output of the style predictor with c as the input image, (σ, Σ) is the multivariate normal distribution with the observed mean and covariance matrix, and α is the hyperparameter augmentation strength, i.e. the ratio of randomness in the style embedding, as opposed to the actual style of the input.

3.2 Loss Function

The overall optimization objective is defined as a combination of the style loss, and content loss [3]:

$$\min_{x} \; \mathcal{L}_c(x, c) + \lambda_s \mathcal{L}_s(x, s) \tag{5}$$

where $\mathcal{L}_c(x, c)$ and $\mathcal{L}_s(x, c)$ are the content loss and style loss components, while λ_s is a scalar hyperparameter which defines the relative weights of the style and content losses.

It is known that higher layers in the network capture the high-level content in terms of objects and their arrangement in the input image, but do not constrain the exact pixel values of the reconstruction. Furthermore, the content loss is defined as an average distance (Frobenius norm) between the high-level semantic features of the content image and the generated image when passed through the pre-trained network:

$$\mathcal{L}_c = \sum_{i \in C} \frac{1}{n_i} \|f_i(x) - f_i(c)\|_F^2 \tag{6}$$

where c and x are the content and restyled images, f is the loss network, $f_i(x)$ is the activation tensor of layer i after passing x through f, and n_i is the number of units in the layer i. Consequently, the style can be represented as a set of Gram matrices that describe the correlations between low-level convolutional features. Therefore, the style loss can be expressed as:

$$\mathcal{L}_s = \sum_{i \in S} \frac{1}{n_i} \| G[f_i(x)] - G[f_i(s)] \|_F^2 \tag{7}$$

where s and x are the style and restyled images, and $G[f_i(s)]$ denotes the Gram matrix of layer i activations of f.

This loss function is using a pretrained VGG19 model, pretrained on ImageNet as a feature extractor for the input and output images. For each input, there are 6 total outputs from the VGG19, 3 of them are low-level and the other 3 are high-level feature maps. Both images are passed through the VGG19 model, followed by calculating the content and style losses (Eqs. 6 and 7) and the total loss, which is a weighted sum of both. Using mini-batch gradient descent for training, the total loss in one forward pass is computed as the average loss out of every sample in the batch. This method does not always result in the fastest model convergence, but it is very computationally efficient.

3.3   Training

The model was trained on a Microsoft Azure Data Science Virtual Machine on Ubuntu, powered by four NVIDIA Tesla K80 graphic cards. The style predictor was trained on 103K images from the Painter by Numbers dataset, while the style transformer was trained on 537K images from the ImageNet dataset. Training was done using mini-batch gradient descent with batch size of 16. The chosen optimizer was Adam, with learning rate = 0.001, beta$_1$ = 0.9, beta$_2$ = 0.999 and no learning decay. Surprisingly, despite the huge computational power of the virtual machine, the training was very slow.

Unfortunately, because of time limitations, the style transformer was trained for a total of only 8 epochs. On the bright side, since the ImageNet dataset is very big, during these 8 epochs the model has observed around 4.3M input images. Because it took so long to train this network, it was impossible to do hyperparameters optimization search, therefore the augmentation strength α was manually tuned. The first 4 epochs had α fixed on 0.45, while the last ones were trained with $\alpha = 0.2$. On every ten thousand steps, the training progress is visualized by generating an image from a random sample (see Fig. 1).

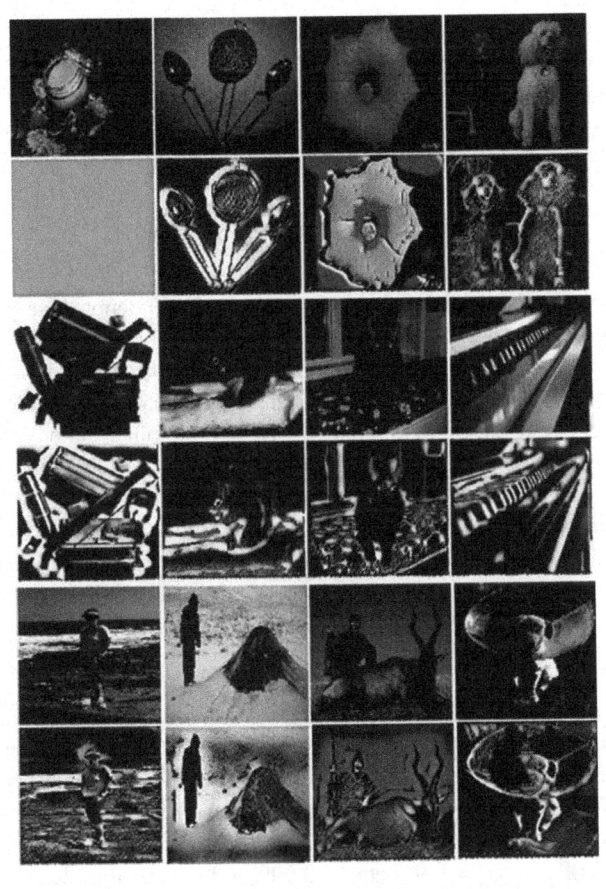

Fig. 1. Training progress of the model, starting from epoch 0 to epoch 7. Odd rows represent the original (input) images, and the even rows represent the corresponding stylized images.

3.4   Experiments and Results

In order to evaluate the augmenter model, we need to check if it is effective in improving the performance of an arbitrary model, by augmenting a dataset the model was trained on, and train it again with the additional data. With the computational limitations we faced, choosing the CIFAR-100 and Tiny-ImageNet-200 datasets to be augmented was an easy choice, since the images are small (32×32 and 64×64 pixels, respectively), so it was easy to compute them in batches. Aside from that, these datasets consisting of 100 and 200 classes respectively are not yet completely solved, as opposed to MNIST or CIFAR-10.

Two Fully Convolutional Networks (FCN) with 142 K parameters were trained, in order to compare the results: one for the original dataset, and another for the dataset combined with its augmented images (with ratio 1:1). While training the evaluator model, none of the other augmentation techniques were used (horizontal/vertical flip, random crop, rotate, zoom, blur, etc.), in order to test the actual contribution of the

method. Training the models on the Tiny-ImageNet-200 dataset lasted 100 epochs. On the other hand, an early stopping method is employed when training both models on CIFAR-100, so the model trained on the original dataset lasted 25 epochs before the validation loss starts to increase, while the latter model lasted 15 epochs (see Figs. 2 and 3). This makes sense because the amount of randomness applied in conditional instance normalization is smaller for low values of augmentation strength (0.45 and 0.2). This means the generated images follow a similar distribution to the input images, which saturate the model quickly.

We can see both models evaluate almost the same on the validation set, and it seems like doubling the CIFAR-100 dataset size using the developed image augmenter did not improve the evaluator model. Similarly, the evaluation on the Tiny-ImageNet-200 dataset shows no improvement on the model (see Fig. 4). The performed experiments showed that the evaluator models did not achieve any significant improvements by using the augmented images in the dataset. However, that may have happened because of multiple factors which are rectifiable, such as:

- The image augmenter was only trained for 8 epochs. This was due to limited computational power and time, but it would be easy to continue learning the model by loading the saved weights.
- No hyperparameters optimization. Most specifically, the augmentation strength may have been too small, and if that is the case, the generated images closely resemble the distribution of the input images, which could mean that the evaluator model did not learn additional significant information from the augmented images. It may be wise to try higher augmentation strength when training the model.
- The CIFAR-100 and Tiny-ImageNet-200 datasets are fully consisting of 32 × 32 and 64 × 64 pixel images, respectively, while the input of the image augmenter is 256 × 256 pixel images. Although nearest neighbor interpolation and reflection fill mode was used, it is possible that some valuable data may have been lost during downscaling and upscaling of the images.

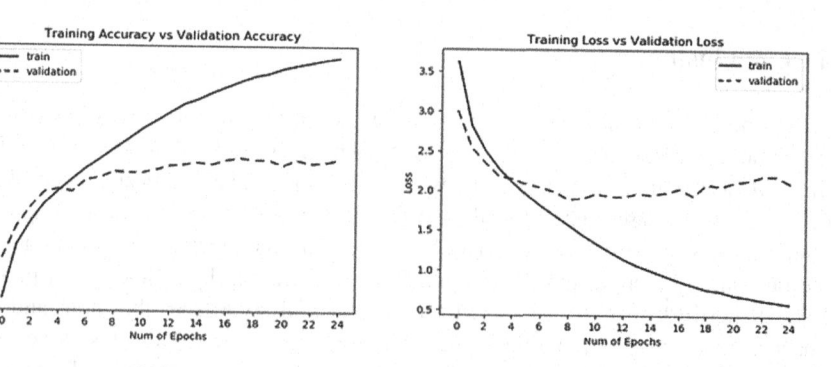

Fig. 2. Train and validation progress of the evaluator through the 25 training epochs, trained only on the original CIFAR-100 dataset.

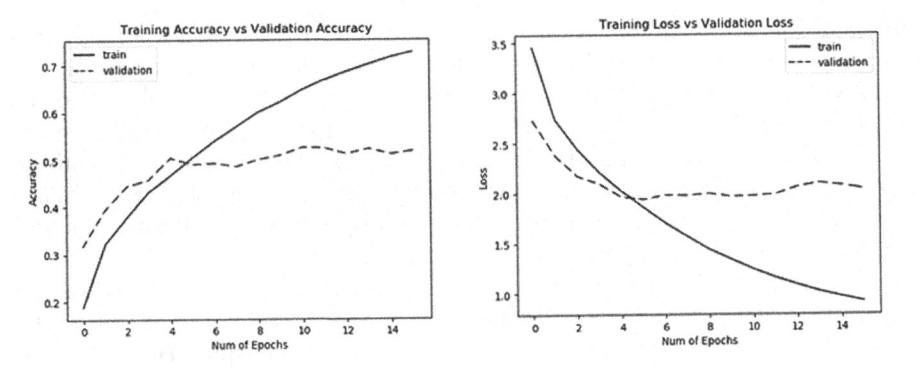

Fig. 3. Train and validation progress of the evaluator through the 15 training epochs, trained on the CIFAR-100 dataset combined with CIFAR-100 augmented images.

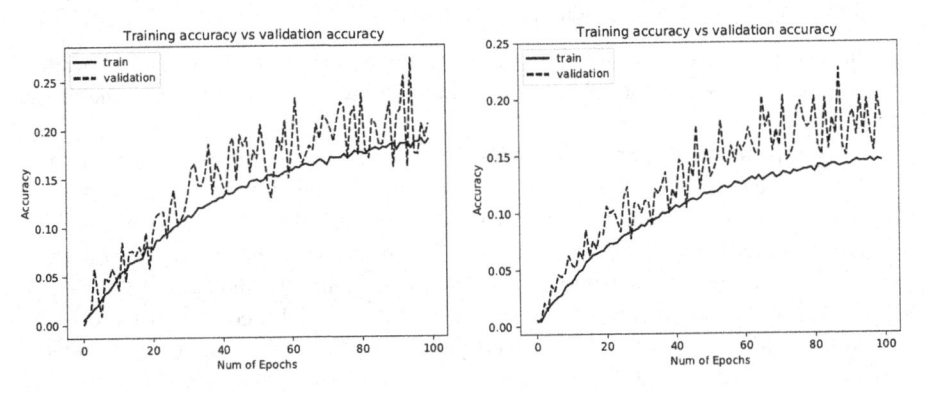

Fig. 4. Comparison between the two models (the left one trained on the original Tiny-ImageNet-200, and the right one trained on a combined dataset) regarding the train and validation accuracy of the evaluator through the 100 training epochs.

4   Conclusion

This paper was an effort to utilize some of the current novel approaches for image generation, with the goal of creating a model which can expand the size of an arbitrary image dataset. Building on top of the work of Dumoulin et al., Ghiasi et al and Jackson et al., two CNNs are trained together to form a complex model capable of applying a randomized style to the input image, by influencing the normalization parameters via the random style embedding. The model is being trained by employing a pretrained VGG19 network in order to extract the content and style from the input and output images. The total loss function is then the combination of content loss (average distance between the high-level semantic features), and style loss (average distance of the Gram matrices between the low-level semantic features). The trained model is later used to augment images from the CIFAR-100 and Tiny-ImageNet-200 datasets, and two identical evaluator models are trained and evaluated for comparing the learnability between the regular datasets and the augmented ones.

The most important conclusion of this paper is that embedding randomized vectors from a previously learned distribution into an encoder-decoder model via conditional instance normalization achieves random stylization in the generated images. By normalizing the feature maps and adding style noise to them, we produce some sort of creativity in the model. This creativity is positively correlated to the augmentation strength α, such that when $\alpha = 0$, the augmenter model should be simply performing the identity function with respect to the input image, while when $\alpha = 1$, the generated image should be random, i.e. its style is 100% drawn from the multivariate normal distribution. Albeit the results were unimpressive, there is much room for improvement. As future work, with enough resources, the potential problems outlined in the previous section can be improved, such as longer training and performing hyperparameter optimization (mostly for the scalar λ_s in the loss function, and the augmentation strength α). Image augmentation with fast neural style transfer seems very promising, and as shown by Jackson et al., such models can be very effective for increasing the size of image datasets.

References

1. Dumoulin, V., Shlens, J., Kudlur, M.: A learned representation for artistic style. arXiv preprint arXiv:1610.07629 (2016)
2. Gatys, L.A., Ecker A.S., Bethge M.: A neural algorithm of artistic style. arXiv preprint arXiv:1508.06576 (2015)
3. Ghiasi, G., Lee, H., Kudlur, M., Dumoulin, V., Shlens, J.: Exploring the structure of a real-time, arbitrary neural artistic stylization network. arXiv preprint arXiv:1705.06830 (2017)
4. He, K., Zhang, X., Ren, S., Sun, J.: Deep residual learning for image recognition. In: Proceedings of the IEEE Conference on Computer Vision and Pattern Recognition, pp. 770–778 (2016)
5. Huang, X., Belongie, S.: Arbitrary style transfer in real-time with adaptive instance normalization. In: Proceedings of the IEEE International Conference on Computer Vision, pp. 1501–1510 (2017)
6. Ioffe, S., Szegedy, C.: Batch normalization: accelerating deep network training by reducing internal covariate shift. arXiv preprint arXiv:1502.03167 (2015)
7. Islam, M., Murase, K., et al.: A new weight freezing method for reducing training time in designing artificial neural networks. In: 2001 IEEE International Conference on Systems, Man and Cybernetics. e-Systems and e-Man for Cybernetics in Cyberspace (Cat. No. 01CH37236), vol. 1, pp. 341–346. IEEE (2001)
8. Jackson, P.T., Atapour-Abarghouei, A., Bonner, S., Breckon, T., Obara, B.: Style augmentation: data augmentation via style randomization. arXiv preprint arXiv:1809.05375 (2018)
9. Kornblith, S., Shlens, J., Le, Q.V.: Do better imagenet models transfer better? In: Proceedings of the IEEE Conference on Computer Vision and Pattern Recognition, pp. 2661–2671 (2019)
10. Krizhevsky, A., Sutskever, I., Hinton, G.E.: Imagenet classification with deep convolutional neural networks. In: Advances in Neural Information Processing Systems, pp. 1097–1105 (2012)
11. Li, J., Gadde, R., Ginsburg, B., Lavrukhin, V.: Training neural speech recognition systems with synthetic speech augmentation. arXiv preprint arXiv:1811.00707 (2018)

12. Simonyan, K., Zisserman, A.: Very deep convolutional networks for large-scale image recognition. arXiv preprint arXiv:1409.1556 (2014)
13. Szegedy, C., Ioffe, S., Vanhoucke, V., Alemi, A.A.: Inception-v4, inception-resnet andthe impact of residual connections on learning. In: Thirty-First AAAI Conference on Artificial Intelligence (2017)
14. Ulyanov, D., Vedaldi, A., Lempitsky, V.: Instance normalization: the missing ingredient for fast stylization. arXiv preprint arXiv:1607.08022 (2016)
15. Yanai, K.: Unseen style transfer based on a conditional fast style transfer network (2017)

Author Index

Printed in the United States
By Bookmasters